ESD Protection Device and Circuit Design for Advanced CMOS Technologies

Oleg Semenov • Hossein Sarbishaei • Manoj Sachdev

ESD Protection Device and Circuit Design for Advanced CMOS Technologies

Authors:

Oleg Semenov
University of Waterloo
Waterloo, ON
Canada

Hossein Sarbishaei
University of Waterloo
Waterloo, ON
Canada

Manoj Sachdev
University of Waterloo
Waterloo, ON
Canada

ISBN 978-1-4020-8300-6 e-ISBN 978-1-4020-8301-3

Library of Congress Control Number: 2008925772

Printed on acid-free paper

9 8 7 6 5 4 3 2 1

springer.com

Dedication

"To my wife, Eva Guerassimova,
for her patience, understanding,
support and encouragement"

Oleg Semenov

"To my parents, Ashraf Moghimi
and Mohsen Sarbisaei"

Hossein Sarbishaei

"To my graduate students"

Manoj Sachdev

Contents

Preface

The challenges associated with the design and implementation of Electrostatic Discharge (ESD) protection circuits become increasingly complex as technology is scaled well into nano-metric regime. One must understand the behavior of semiconductor devices under very high current densities, high temperature transients in order to surmount the nano-meter ESD challenge. As a consequence, the quest for suitable ESD solution in a given technology must start from the device level. Traditional approaches of ESD design may not be adequate as the ESD damages occur at successively lower voltages in nano-metric dimensions.

This book makes an attempt to address ESD circuit design issues in a systematic manner. As ESD event is a high current/voltage phenomenon, predicting device behavior in this regime is only possible with device-level simulators. Yet, in a modern CMOS technology most of the process parameters are not available to majority of designers, especially in fabless companies. Therefore, the device parameters should be first calibrated with the given process technology at normal operating conditions. In the next step these process parameters are used to create and simulate the individual ESD protection devices in the device simulation environment. After successful design of an ESD protection element, mixed-mode, circuit-device level simulations are carried out to ensure proper ESD circuit design for a given purpose, say I/O or a clamp. Finally, a test chip is fabricated to evaluate performance of the ESD protection circuit with standard ESD measurement equipments, i.e. HBM/MM/CDM testers and TLP tester.

We started to investigate the ESD circuit design issues at University of Waterloo in 2001 for high speed I/Os. Very soon it became apparent that

ESD circuit design and layout is an art rather than a science. Our objective was to establish a design flow so that interested ESD designers could use it for robust ESD protection circuit design. Furthermore, we were also interested in establishing a cause and effect relationship between design parameters and achieved ESD protection early in the design phase in order to reduce the design time. Over the last several years, we designed a variety of ESD protection circuits in different technologies following the above mentioned design flow. They are described in various chapters of the book.

This book is intended for engineers working in the areas of device and I/O circuit design. In addition, as the problems associated with ESD failures and yield losses become significant in the modern semiconductor industry, the demand for graduates with a basic knowledge of ESD is also increasing. Today, there is a significant demand to educate the circuits design and reliability teams in ESD, since nuisance value of inadequate ESD protection is very high. It should be noted that a large volume of work has done on various aspects of ESD and is scattered in publications. Authors acknowledge that there are few very good quality books on the subject. We have made an effort to optimize ESD as well as circuit design objectives. In addition, we have tried to cover some of the recent topics. Some of the highlights of the book are listed below:

(i) The charge board ESD (CBM) testing that is becoming popular in robust PCB designs used in wireless products such as cell phones and PDAs.

(ii) The impact of burn-in testing (accelerated test methods) on the ESD robustness of deep sub-micron ICs.

(iii) Distributed ESD protection networks optimized for sub-90 nm CMOS ICs.

(iv) ESD protection strategies for smart power ICs that are widely used in automotive industry.

ESD professionals have a several challenges to face as we further scale into nano-metric regime. Due to technology scaling and expansion of automated handling, failures in ICs caused by Charged Device Model (CDM) ESD are an increasingly important reliability issue. Even today, a significant portion of ESD field returns is due to damages originating from CDM stress. Moreover, CDM discharges can cause latent damages which could degrade and eventually lead to definite failures in the ICs. The ESD protection design for current and future sub-65 nm CMOS circuits is a challenge for high I/O count, multiple power domains and flip-chip products. For example, 90 nm

CMOS ASIC design systems offer over 1,500 I/Os and more than 200 analog and high speed serial I/Os. As a consequence, the development and testing of a new sub-65 nm CMOS ESD protection circuits will become a crucial task for academia and semiconductor industry.

Acknowledgments

Several individuals, and organizations have contributed in one way or the other to further the objectives of our ESD research presented in this book. Authors would like to acknowledge several discussions with Vasilis Papanikolaou, Efim Roubakha, Manuel Palacios and Andrew Neely from Gennum Corporation on analog circuits design, ESD protection concepts and ESD measurements.

Authors gratefully acknowledge the National Sciences and Engineering Research Council of Canada (NSERC) and Gennum Corporation for the financial support of the ESD research. The results of this research became an important part of this book. We make special acknowledgement to Dr. Valery Axelrad from the Sequoia Design Systems, Inc. for the providing us the ESD TCAD simulation tool and fruitful discussion of simulation results. The authors would like to thank all members of CMOS Design and Reliability Group of the University of Waterloo for their important and vital contribution to this book. We would also like to thank the Senior Publishing Editor, Dr. Mark de Jongh, for good and supportive collaboration.

Chapter 1

INTRODUCTION

1. NATURE OF ESD PHENOMENA

One of the most observed sources of electrostatic charge is the shock caused by touching a doorknob after walking in a carpeted room. This shock is a result of discharging the body's accumulated charge through a conductive object. Normally, this electrostatic discharge can be a few kilo-volts. In semiconductor industry the discharge path can be through semiconductor devices. Before going through the details of this phenomenon in semiconductor devices, a brief review of static electricity is necessary.

Static electricity is the creation of electrical charge by an imbalance of electrons on the surface of a material which produces an electrical field. When two objects with different electrical potentials are brought in contact, a charge transfer occurs between these two objects. This phenomenon is called electrostatic discharge. There are different ways to create a charge on a material: triboelectric charging, induction, ion bombardment and contact with another charged object. The most common mechanism is triboelectric charging. Triboelectric charging is the creation of charge by the contact and separation of two materials. Consider contact and separation of two uncharged materials. As a result, based on the nature of materials, electrons transfer from one material to the other. Therefore, material that looses electrons becomes positively charged while the other material becomes negatively charged. The amount of this triboelectric charge depends on many factors such as area of contact, speed of separation and relative humidity. Table 1-1 shows some examples of the amount of generated electrostatic charge under different conditions and for two different relative humidity

O. Semenov et al., ESD Protection Device and Circuit Design for Advanced CMOS Technologies, 1–19.

situations. It can be seen that higher humidity reduces the generated charge significantly. Electrostatic discharge occurs when this charge is transferred to another material. The resistance of the actual discharge circuit and the contact resistance at the interface between contacting surfaces determines the charge that can cause damage.

As mentioned earlier, the polarity and magnitude of the electrostatic charge depends on the material's characteristic. A table called triboelectric

Table 1-1. Examples of generated electrostatic charge.

Means of generation	10–25% Relative humidity (kV)	65–90% Relative humidity (kV)
Walking across carpet	35	1.5
Walking across vinyl tile	12	0.25
Worker at bench	6	0.1
Poly bag picked up from bench	20	1.2
Chair with urethane foam	18	1.5

Table 1-2. A typical triboelectric series.

Material	Electrostatic polarity
Air	
Human skin	Most positive (+)
Glass	
Human hair	
Nylon	
Wool	
Silk	
Aluminum	
Paper	
Cotton	
Steel	
Wool	
Hard rubber	
Nickel, copper	
Brass, silver	
Gold, platinum	
Acetate fiber (rayon)	
Polyester (mylar)	
Celluloid	
Orlon	
Polystyrene (styrofoam)	
Polyurethane	
Saran	
Polyvinyl chloride	
Silicon	
Teflon	Most negative (–)
Silicon rubber	

series classifies different materials based on their electrostatic property. Table 1-2 shows the triboelectric series table for different materials. It can be seen that air and human skin are capable of carrying the most positive charge, while silicon rubber, teflon and silicon are capable of carrying the most negative charge.

In a triboelectric charging event, the object that is closer to the top of the table takes a positive charge and the other one takes a negative charge. In addition, materials that are further apart on the table generate higher charge than those that are closer. Based on the triboelectric series table, all materials can be electrically charged. However, the amount of generated charge and where and how fast the charge goes depend on the material's electrical characteristics. Insulators, due to their very high resistance, are capable of storing a huge amount of electrostatic charge. As this charge cannot move, it remains on the surface of the material. On the other hand, when an inductor is charged, due to its very low resistance, the generated charge uniformly distributes across the surface of the material.

2. ESD FAILURES IN NANOMETRIC TECHNOLOGIES

Electrostatic Discharge (ESD) is a common phenomenon in the nature. As mentioned in the previous section, the amount of electrostatic charge can be a few kilovolts, which is being discharged extremely fast (in the order of tens of nanoseconds). ESD is a subset of a class of failures known as electrical overstress (EOS). The EOS class is composed of events that apply conditions outside the designed operating environment of the part. These conditions include voltage, current, and temperature.

ESD events occur throughout a product's life. ESD first affects the integrated circuit (IC) early in the wafer-fabrication process. Clean-rooms are good sources for charge-generating materials due to the extensive use of synthetic materials in containers and tools [1]. The electrostatic discharge may reduce the product yield in two ways. A photolithography operation transfers the mask image to the silicon wafers. The ESD event distorts the fine pattern defined on the mask. If the mask is damaged by ESD, then each circuit is printed with this damage [2]. The second mode of ESD damage is a direct discharge to the wafer. This event results in a gate oxide and junction damages [3].

The next stage in a product's life is the assembly operation, where additional ESD hazards are present. In the assembly operation, the wafers must be cut to yield individual dies. These dies are then inspected and placed in packages. Wires are attached to allow signals to travel from the outside pins to the die. Finally, the package is formed around the die. All of these operations are capable of producing ESD events [4].

As microelectronics technology continues shrink to nano-metric dimensions, ESD damage in integrated circuits has become one of the major reliability issues. Several studies carried out over the past two decades ranked ESD and EOS (electrical overstress) as the major cause for field returned ICs, as it shown in Table 1-3. It was found that ESD related failures can reach up to about 70% failure modes, depending on the product type [5].

Table 1-3. The EOS/ESD failure percentage as total failure modes studied by different author over the years. (Adapted from [6].)

Percentage	51%	72%	23%	38%	45%	43%	39%
Author	Green	McAteer	Euzent	Merill	Wagner	Shumway	Brodbeck
(year)	(88)	(88)	(91)	(93)	(93)	(95)	(97)

The thinner gate oxide and shallower junction depth used in the advanced technologies make them very vulnerable to ESD damages. The silicidation reduces the ballast resistance provided by drain contact to gate edge spacing (DCGS) with at least a factor of 10. As a result, scaling of the ESD performance with device width is lost and even zero ESD performance is reported for standard silicided devices [6].

ESD failures are caused by at least one of three sources: localized heat generation, high current densities, and high electric field intensities. The current densities associated with an ESD stress unavoidably imply high power dissipations, with consequent rise in the lattice temperature that often results in thermal damages. Silicon has a negative resistance relationship with temperature coefficient, so very high power dissipation in a small volume will result in higher temperatures and thermal runaway. As long as the rate of heat removal is greater or equal to the rate of heat generation, the junction temperature does not increase. When the rate of heat generation becomes greater than the rate of heat removal, the junction temperature in hot spot region starts to increase and thermal runaway occurs.

For CMOS circuits, the electric field intensity refers to the voltage developed across the dielectric and junctions in the circuit. The gate oxide is the most vulnerable dielectric owing to its thinness. Structural defects and sharp corners in layouts result in higher electric field and current crowding making failure more likely at these points.

ESD induced failures can be grouped in two categories: soft and hard failures. In case of soft failure, the device has a partial damage typically resulting in an increased leakage current that might not meet the requirements for a given circuit. Still, the basic functionalities of the device are operative but without any guarantee about potential latency effects. In case of hard failures, the basic functionalities of the device are completely destroyed during the ESD event. Each ESD failure mode is traced to one or more of four fundamental damage mechanisms [7].

2.1 Oxide Rupture (Breakdown)

Typically, gate oxides can withstand an electric field of 6–10 MV/cm before it breaks down. In CMOS technology input/output buffers require an ESD protection circuit that limits the peak voltage during an ESD event that could cause irreversible failure (rupture) of the gate oxide. Being the peak voltage (V_{peak}) the maximum voltage across the protected device, it is necessary to maintain a margin between this voltage and the gate oxide breakdown (V_{BD}) to avoid oxide failures. The gate oxide breakdown (V_{BD}) is a critical function of its thickness. But with the scaling down of the device sizes, the gate oxide thickness is also reduced resulting in a decrease of the V_{BD}. Figure 1-1 shows the gate oxide damage in a MOS transistor after the CDM stress.

This defect results in a low-ohmic short of gate and drain terminals in damaged transistor [8]. The gate oxide is not the only concern in integrated circuits. Both bipolar and MOSFET processes can have dielectric ruptures in oxides over active circuitry. This may be the oxide grown over a diffused resistor or an isolation region. If a conductor passes on top of this oxide, a rupture can occur.

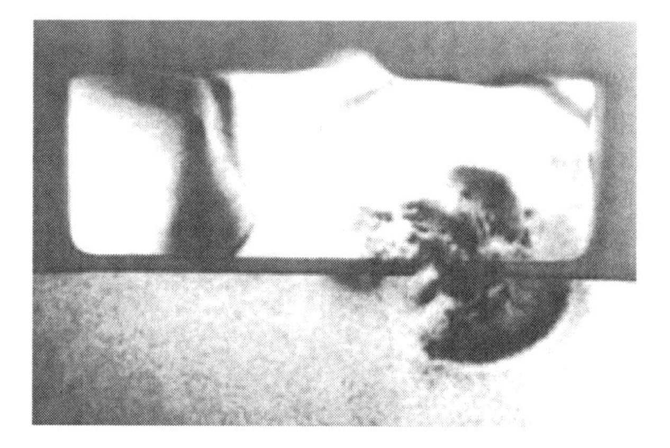

Figure 1-1. Gate oxide damage in MOS transistor after the CDM stress.

2.2 Junction Filamentation and Spiking

Junction filamentation causes an increase in the reverse bias leakage of a *p-n* junction. In the worst case, the junction is shorted. The ESD event causes current to flow through the junction. The high power dissipated in the junction cause the temperature to rise until a region of silicon melts. When silicon melts, its resistance drops by a factor of 30 or more [9]. This causes more of the current to flow in the narrow, melted region which further heats the melted region, leading to thermal runaway. This phenomenon is often referred to as the second or thermal breakdown [10]. In MOSFET devices the drain junction filamentation is typically located close to the surface in the gate to drain overlap region, where the dielectric acts a thermal insulator, as it shown in Figure 1-2.

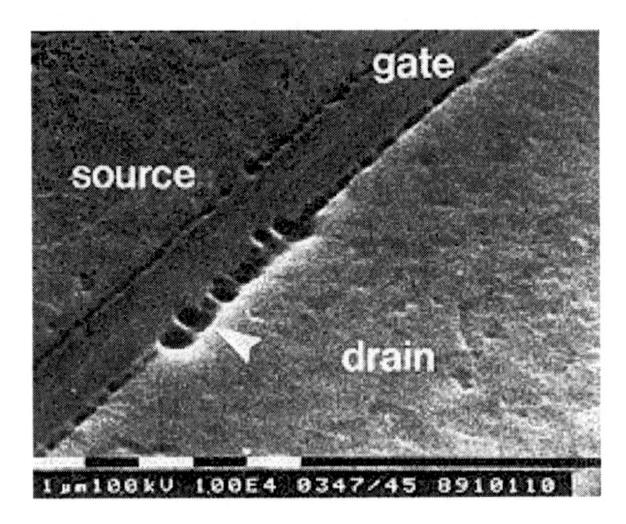

Figure 1-2. Drain junction filamentation in MOS transistor due to the ESD stress [11].

Therefore, devices, in which hot spot regions are located deeper in the silicon (like BJT, SCR and Thick Field Oxide Device), are used as a high reliable ESD protection devices. In bipolar junction transistors (BJTs) the base-emitter junction is the most susceptible to the filamentation damage. Junction spiking is a similar mechanism, except the melted region grows until it intercepted by a metal contact, causing the degradation of aluminum and silicon. The damage thresholds are lower for aluminum contacts because the aluminum-silicon eutectic forms at 577°C rather than at the melting point of silicon (1,415°C) [12].

2.3 Metallization and Polysilicon Burn-out

Thin-film fusing affects each conducting film in a circuit. These include the metal interconnects, polysilicon interconnect, and thin-film and diffused resistors. The most susceptible to damage are circuits with thin-film resistors. It is important during the chip design that the resistor be made wide enough to handle an ESD current pulse for the desired level of protection. Because of the high temperatures induced by the ESD pulse, a metal or polysilicon line, located close to the hot spot region in *p-n* junction, can be melted resulting in a metal opens, as it shown in Figure 1-3.

Figure 1-3. Interconnects damage due to the ESD stress [13].

2.4 Charge Injection

The last degradation mechanism is the charge injection into the gate oxide by avalanche breakdown of *p-n* junction. This occurs when an ESD event causes a reverse-biased junction to conduct by avalanche multiplication. Some of the carriers have enough energy to surmount the oxide-silicon energy barrier, as it is illustrated in Figure 1-4. If the junction is the drain of a MOSFET, it results a shift in the threshold voltage of transistor [14]. The degree of degradation in oxide reliability is also related to the current density of the injected charge as well as the total charge injected [15]. A localized charge injection during the drain junction avalanche breakdown at the ESD event causes more damage than the uniform charge injection from the gate to the body of a transistor at normal operating conditions. The charge injection induces the increasing of leakage current in the damaged devices.

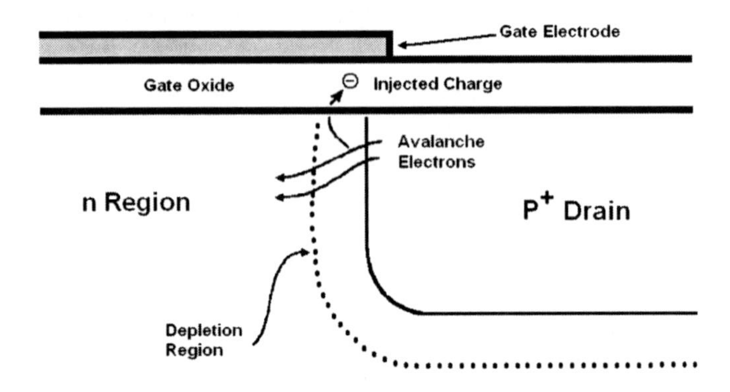

Figure 1-4. Mechanism of charge injection in gate oxide region under ESD stress [7].

3. CIRCUIT RELIABILITY: ESD MODELS

The most challenging aspect for ESD reliability is the protection circuit effectiveness or ESD protection level, especially in highly susceptible advanced CMOS technologies. Higher current densities and reduced ballasting resistances favor localization of stress currents and degrade the ESD performance of I/O protection devices in advanced CMOS processes with shallow Source/Drain junctions and silicided diffusions. Currently, the most commonly used ESD models for the reliability estimation of VLSI ICs are: Human Body Model (HBM), Machine Model (MM) and Charge Device Model (CDM). The basic model for ESD protection is the HBM intended to represent the ESD caused by human handling of ICs. Typically, HBM events occur at 2–4 kV in the field, hence, protection levels of this range are necessary. Besides human handling, contact with machines is also an ESD-type stress event. Since the body resistance is not involved here, the stress is severe with relatively higher current levels, thus, protection levels of 200 V for this model usually ensure adequate device reliability. The CDM ESD is intended to model the discharge of a packaged IC. Charges can be placed on an IC either during the assembly process or on the shipping tubes [16]. The protection level of 500 V for CDM ESD stress is typically used to ensure the chip reliability. The details of reliability ESD models are discussed in Chapter 2.

4. ESD CHALLENGES FOR ADVANCED CMOS TECHNOLOGIES

Semiconductor technology evolution results in technology scaling of geometric dimensions and lower power supplies. It introduces a new materials, features, devices and integration. Evolutionary changes in CMOS technology included the moving from local oxidation (LOCOS) based isolation to shallow trench isolation (STI), from diffused n-well to retrograde implanted wells, from non-salicide to salicide junctions, etc. In addition, the recent technology scaling from 180 nm CMOS to 65 nm CMOS resulted in moving from aluminum (Al) to copper (Cu) metallization and from silicon dioxide to low-k materials for inter-level dielectrics (ILD). All of these changing impact the ESD robustness of semiconductor products with both positive and negative ramifications.

4.1 Scaling of Metal Interconnects: Aluminum to Copper

The trend in advanced CMOS technology generations is a migration to interconnect systems using the dual damascene Cu-based metallization, exhibiting an improvement in the electrical conductivity compared to Al based interconnects with the same effective line thicknesses. The need for reduced resistance and capacitance interconnects has led to an evolution from Ti/Al/Ti based technology to Cu-based interconnect processes, since the Cu wiring has 35% lower resistance, higher allowed current density and 20% lower cost compared to Al-based interconnects [17].

As metal lines decrease in thickness with each technology generation they are becoming "fuses" in the Electrostatic Discharge (ESD) protection path if not sized correctly. The continuous scaling down of ICs for faster switching speed has decreased the size of both the devices and the metal lines that form interconnects between the devices and power rails in a chip. This reduction in the dimensions of interconnects affects their ESD robustness. It is therefore important to characterize the heating effects of copper lines under ESD conditions and set guidelines consistent with the wire current carrying capability to handle an ESD event. During an ESD stress, metal lines fail if Joule heating results in raising the conductor temperature beyond its critical temperature (T_{crit}). In addition, an ESD induced electromigration may also cause the conductor failure [17, 18]. Figure 1-5 shows a cross section of a Ti-clad aluminum and cladded dual-damascene Cu-based interconnects placed in a polymeric dielectric insulator [19].

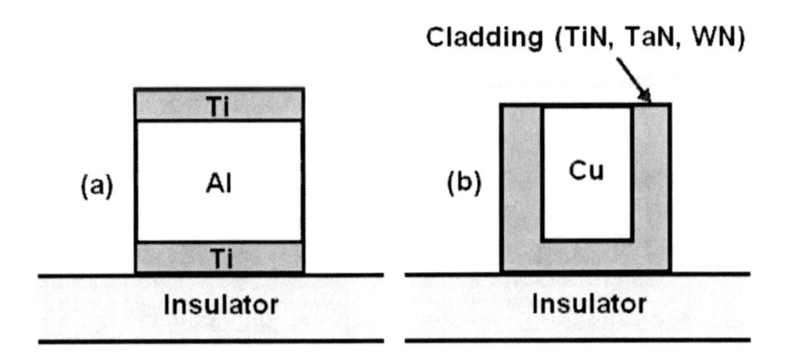

Figure 1-5. Cross section of Al (a) and Cu (b) interconnect with cladding material. (Adapted from [17].)

In the damascene process, the polyimide film is etched, followed by cladding material and Cu film deposition. The liner cladding material can be TiN, WN, Ta, TaN, or TaSiN to form good adhesion and act as a copper diffusion barrier. Copper interconnects are geometrically different from the Ti/Al/Ti interconnects in that the cladding is on the bottom and the two sides. In a dual-damascene Cu-based fabrication process, Cu-vias are filled concurrently with the Cu-interconnects by creating a trough in the polyimide with two different depths.

The ESD robustness of Cu-based interconnects has been shown to be two or three times superior to Al-based interconnects (for wires and vias, respectively) based on transmission-line-pulse (TLP), HBM, and MM testing

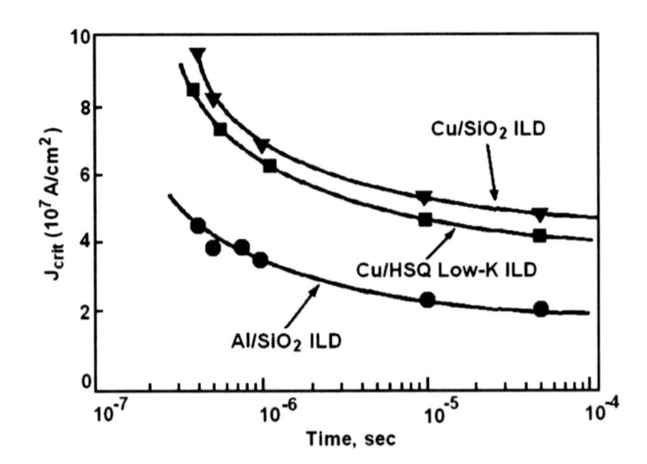

Figure 1-6. Jcrit as a function of ESD pulse width for copper and aluminum interconnects with different ILD materials: SiO_2 and low-k hydrogen silsesquioxane (HSQ) [20].

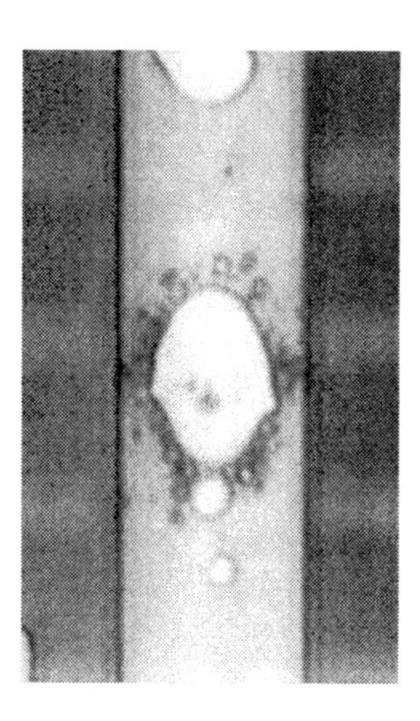

Figure 1-7. ESD induced failure of a copper interconnect [21].

(Figure 1-6) [20]. By migrating to Cu-based interconnects, the critical-current-density to failure (J_{crit}) and ESD HBM and MM improvements allow to continue scaling of metal film thickness and line width, which improves *rc* delay and ESD reliability. The low resistivity, high melting point and good mechanical strength explain the better ESD performance of Cu interconnects than interconnects implemented from conventional materials such as Al or AlCu alloy. Note, that Cu melting temperature is 1,034°C in comparison to an Al melting temperature of 660°C [17]. Al-clad and Cu-clad interconnect failures are different because of the geometry and materials in the interconnect structure. In Al-based interconnects, the cladding is on top and bottom. In this case, the dielectric cracking and extrusion occurs laterally. In Cu-based interconnect systems, cladding is on the two sides and bottom; dielectric cracking occurs vertically and weak extrusion occurs laterally. Figure 1-7 shows an example of the onset of ESD-induced failure of a Cu interconnect film. The peak temperature occurs in the center of the wire leading to the onset of failure in the center region [21].

4.2 Scaling of Inter-Level Dielectrics: SiO_2 to Low-k Materials

Low-k inter-level dielectric (ILD) materials further reduce interconnect capacitance. The International Technology Roadmap for Semiconductors (ITRS) states that ILD electrical permittivity must be scaled from k = 4.0 to k = 1.5 to meet signal delay requirements for sub-0.18 μm CMOS technologies [22]. From an ESD perspective, the thermal properties of low-k ILD materials are key to understanding the ESD robustness of Cu interconnect systems and the scaling implications on ESD robustness. Low-k materials impact ESD robustness of interconnects because of the decreased thermal conductivity and increased thermal impedance in comparison with traditional SiO_2. The results, presented by IBM, shown that the thermal impedance, θ_{th}, of Cu interconnects in low-k hydrogen silsesquioxane (HSQ) ILD (k = 2.9) increased 12%, and 35% for M2 and M1 interconnects, respectively [20]. As depicted in Figure 1-6, the high-current TLP testing of dual-damascene Cu wires in low-k HSQ ILD shows a 15% decrease of J_{crit} (the critical current density to failure) compared to Cu wires in SiO_2 ILD. The results of HBM ESD testing of Cu interconnects in SiO_2 and low-k ILD as a function of Cu interconnects line-width are shown in Figure 1-8 [23].

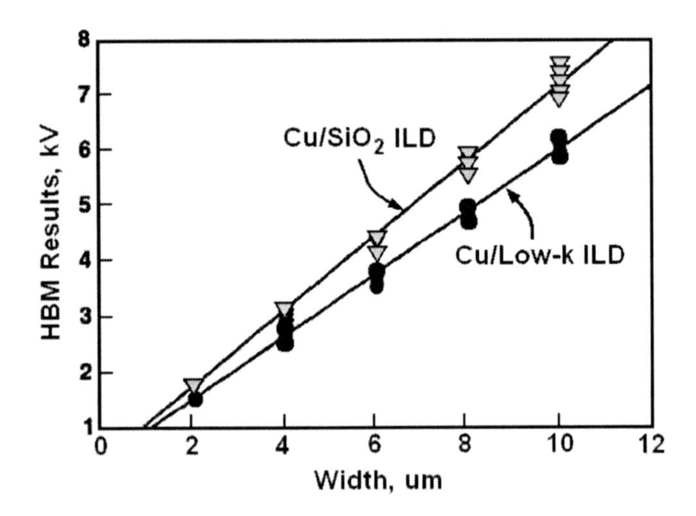

Figure 1-8. HBM ESD failure dependence of Cu interconnects in both SiO_2 and low-k HSQ dielectric versus Cu line widths. (Adapted from [23].)

From this figure we can conclude that the Cu interconnect in the SiO_2 ILD interconnect system is superior to the identical Cu interconnect in low-k ILD. For a given applied current pulse, Cu interconnects in low-k ILD reach a higher temperature compared to the Cu interconnect in SiO_2 ILD.

Because of the different geometrical interconnect structures of the line, cladding and insulator material, failure analysis of Cu interconnects in SiO_2 ILD after HBM ESD stress testing shows unique failure mechanisms not typically observed in Al-based interconnects. In Cu damascene interconnects, ESD failure mechanisms demonstrate cracking of the dielectric above the film in passivated structures. In an unpassivated damascene Cu interconnects, the displacement, hillocking and blistering were observed after HBM ESD testing. CDM pulses are significantly faster than MM and HBM pulses and the Joule heating does not have a significant time to penetrate into the ILD volume. No insulator degradation was observed under CDM ESD events [23]. Cu interconnects are typically failed before ILD.

5. ESD DESIGN WINDOW

Many different aspects need to be taken into account for the development of ESD protection concepts in nano-metric CMOS technologies. It starts with the selection of suitable protection elements and ends in the choice of an ESD optimized circuitry. All ESD devices have to fulfill certain conditions concerning their I-V characteristics, which are described by the ESD design window shown in Figure 1-9 [24]. In addition, layout and process technology issues determine the clamping and shunting capabilities of the protecttion devices.

The protection device should not limit the normal operation, i.e. its triggering voltage (V_{t1}) must be above the signal range plus some noise margin. Under normal operating conditions for the IC, the ESD device must be in the

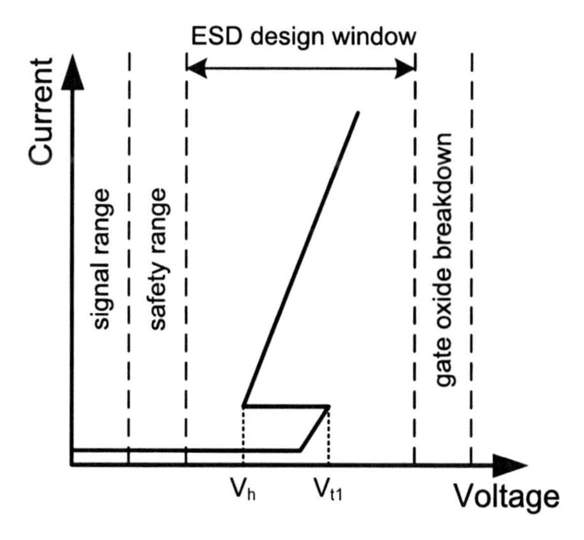

Figure 1-9. ESD design window between the maximum operating voltage and the gate oxide breakdown voltage of the core circuit (Vh is the holding voltage, Vt1 is the triggering voltage.). (Adapted from [24].)

off-state. On the other hand, the voltage drop across the protection element at the ESD relevant current densities of several 10 mA/μm should not exceed a critical limit which could lead to a damage of the protected circuit like a gate oxide of input/output buffers and core transistors. Obviously, the protection device itself should withstand a high amount of discharge current which can be in the range of several amps for several 100 ns of ESD pulse.

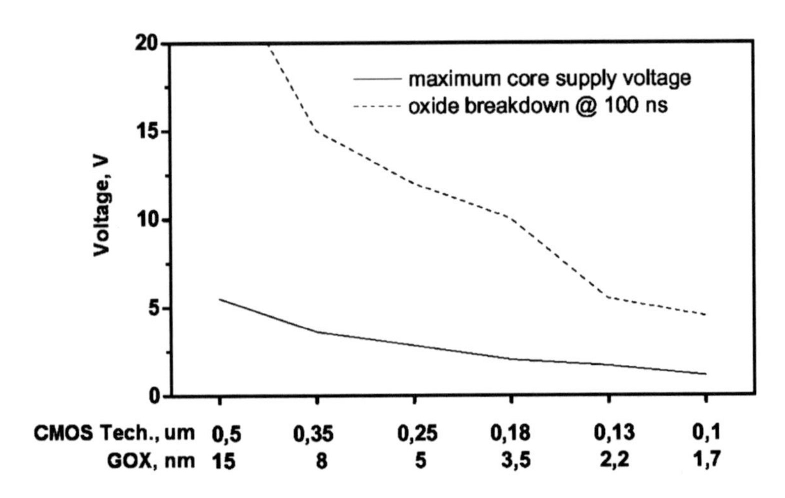

Figure 1-10. Breakdown voltage of gate oxide for 100 ns duration stress pulses and supply voltage as a function of CMOS technology generation. (Adapted from [25].)

Figure 1-10 illustrates the maximum core supply voltage as well as the oxide breakdown voltage for different technologies [25]. These data were obtained under ESD conditions with pulse duration of 100 ns. As it is apparent from the figure the gate oxide breakdown voltage is reduced significantly while the core supply voltage is not reduced aggressively. As a consequence, the width of the ESD design window for the core protection is reduced to about 3 V for the 100 nm node which must be guaranteed for the all process corners and the full operational temperature range. This requires a sophisticated balancing of the protection device features and their careful analysis including process fluctuations.

6. BOOK OBJECTIVE AND ORGANIZATION

ESD protection has been an essential component of integrated circuits ever since the invention of semiconductor devices. A significant number of protection devices have been designed and optimized for this purpose by ESD engineers. However, the ESD phenomena in VLSIs is typically outside of the usual operating conditions of the devices being stressed. Thus, the behavior of semiconductor circuit elements during an ESD event is not covered by standard text books on semiconductor device physics and VLSI design. In the sub 100 nm CMOS technologies, the ESD protection development must go beyond the development of a specific optimized protection element. A sophisticated protection network has to be designed, which covers both the IO circuits and the core region, where lower oxides thickness and low junction breakdown voltages lead to hard constraints on the maximum voltage overshoot during the ESD event. In addition, the circuit designs with multiple power supply domains will further complicate the ESD supply protection concepts. In this book we have made a conscious attempt to enable the reader to gain a broad understanding of ESD events in deep submicron VLSIs; the general issues associated with improving of ESD performance and robustness. The book draws from a large journal and conferences publication base and authors' extensive research experience in the field of ESD protection area. It consists on nine chapters and the brief outline of the contents of each chapter is given below.

Chapter 2 presents a detailed review of ESD phenomena, the appropriate test methods, and ESD models which are widely used in semiconductor industry. An ESD event may carry amperes of current in a short period of time ranging typically from hundreds of pico-second to hundreds of nanosecond. For the purposes of reproduction under controlled conditions, the real word ESD events are classifieds in four main categories under which electronic elements or ICs are tested: the human body model (HBM), the machine model (MM), the charge device model (CDM) and the printed circuit board (PCB) ESD model. The issues dealing with the accuracy of these models are also discussed in this chapter.

To understand the mechanisms of device failure and the operation of deep submicron protection devices under the high current conditions and short duration of ESD pulses, one must understand the basic device physics behind these concepts. These issues are considered in Chapter 3. Various semiconductor devices can be used as an ESD protection circuit. Generally, based on the shape of the I-V characteristic of semiconductor devices, they are divided into two main categories: non-snapback devices and snapback devices. In this chapter, some of the most important devices that are used in CMOS ESD protection circuits are discussed. In addition, special accelerated test methods such as burn-in are often employed as reliability screens to weed out infant mortalities. Weak gate oxides are one of the major components of such failures. These failures are accelerated due to elevated temperature (~125°C), elevated voltage (V_{DD} + 30%) and long stress time

(30–168 h). Under stress operating conditions, ESD/EOS robustness of protection devises becomes worse. The robustness of ESD devices under stress conditions is also discussed in this chapter.

In Chapter 4, the design requirements for effective protection circuit that can perform without degrading of VLSI ICs functions are discussed. For example, a protection at the input should not affect the gate oxide reliability of core transistors, an output protection should have no impact on the output buffer performance and neither of these should result in increase in the leakage current in the chip. At the beginning of this chapter the different ESD protection networks for different applications are considered. The second part of the chapter describes the design flow of ESD protection network. It covers the ESD device simulations and calibration, mixed-mode (device-circuit) simulations and chip level ESD simulations. The design of special test structures for ESD network verification and ESD measurement concepts are also considered in this chapter.

Whole-chip ESD protection has become an important reliability issue of CMOS ICs. Even if there are suitable ESD protection circuits around the input and output pads, the internal circuits are still vulnerable to the ESD damage. Thus, an effective ESD clamp circuit between the power rails is necessary for the protecting of internal circuits against the ESD damage. ESD clamps can come in many different varieties, as it is shown in Chapter 5. Generally, ESD clamps can be grouped into two categories: static and transient. The static clamps provide a static or steady-state current and voltage response. A fixed voltage level activates static clamps. As long as the voltage is above this level, the clamp will conduct current. A diode, MOSFET and SCR based clamps are typically used as static ESD clamps. Transient clamps take advantage of the rapid changes in voltage and/or current that accompanies an ESD event. During this transient, an element is turned on very quickly and slowly turns off. This type of clamp conducts for a fixed time when it is triggered. An RC network determines the time constant. These clamps are typically triggered by very fast events on the supply lines.

The performance degradation of high-speed differential I/Os, which are widely used for high-speed mixed mode applications, due to the ESD protection network is discussed in Chapter 6. The layout optimization issues, which can reduce the parasitics of ESD protection devices, are also considered in this chapter.

In compact microelectronic systems, it is necessary to combine different functional blocks on one chip to minimize the component count. This smart power approach enables designs with a mixture of CMOS logic, low voltage analogue and high voltage driven transistors on one chip to realize System on Chip (SoC). In Chapter 7, the ESD protection strategies of high voltage modules in smart power ICs are discussed. There are two general categories of ESD protection schemes used in I/Os: the non-self protecting scheme and the self protecting scheme. For non-self protecting scheme, it needs to add the ESD protection devices to the I/O cell. While for the self-protecting scheme, the I/O cell itself is an ESD protection device. In smart power technologies, the typical self protecting device is the lateral diffused MOS (LDMOS) power transistor since the device can be used as the ESD protection device and as output driver simultaneously. The typical non-self protecting scheme for power technologies is based on ESD high-voltage MOSFETs, silicon controlled rectifier (SCR) devices and bipolar junction transistors (BJTs). This chapter is focused on the analyzing of ESD robustness of different power ESD protection devices used for smart power applications, its latch-up immunity and layout issues.

ESD related issues of RF CMOS circuits are covered in Chapter 8. Nowadays, the design of modern high-performance high speed and analog RF circuits leaves very small design window for direct implementation of the ESD protection elements due to the high sensitivity of the core RF circuits to even a small parasitics and the generally increased device susceptibility to ESD stress with the CMOS technology downscaling. The most critical components of narrowband transceiver front-ends are the low noise amplifier (LNA) and the power amplifier (PA) due to the high requirements on their RF performance, decreasing further the already tight ESD design window. The ESD protection strategies used for these RF circuits are discussed in this chapter.

Finally, in Chapter 9 the main topics covered in this book are summarized. This summary is followed by the main challenges in ESD protection for advanced technologies along with some directions on the required future works in this area.

7. SUMMARY

Electrostatic discharge (ESD) is a subclass of the failure known as the electrical overstress (EOS). The ESD event has four major stages: (1) charge generation; (2) charge transfer; (3) charge conduction; and (4) charge-induced damage. ESD protection strategy has to minimize the charge generation and slow the charge transfer by controlling the environment where parts are handled and stored. The next aspect focuses on the circuit elements. Here, protection techniques should make the individual elements more robust to the high current conditions by adding additional ESD protection devices to change the conduction paths through a circuit. In this chapter, the typical ESD induced failures were considered and the ESD design challenges for advanced CMOS technologies were discussed. The book objectives and its organization were also presented in this chapter.

REFERENCES

[1] S. U. Kim, "ESD induced gate oxide damage during wafer fabrication process," *EOS/ESD Symposium*, pp. 99–105, 1992.

[2] R. G. Chemelli, B. A. Unger, and P. R. Bossard, "ESD by static induction," *EOS/ESD Symposium*, pp. 29–36, 1983.

[3] C. Diaz, S. M. Kang, and C. Duvvury, "Tutorial electrical overstress and electrostatic discharge," *IEEE Trans. on Reliability*, vol. 44, No. 1, pp. 2–5, 1995.

[4] J. Bernier and G. Groft, "Die level CDM testing duplicates assembly operation failures," *EOS/ESD Symposium*, pp. 117–122, 1986.

[5] C. Russ, ESD Protection Devices for CMOS Technologies: Processing Impact, Modeling, and Testing Issues, Ph.D. thesis, 1999.

[6] J. -B. Huang and G. Wang, "ESD protection design for advanced CMOS," *Proc. of SPIE*, vol. 4600, pp. 123–131, 2001.

[7] J. E. Vinson and J. J. Liou, "Electrostatic discharge in semiconductor devices: protection techniques," *Proc. of the IEEE*, vol. 88, No. 12, pp. 1878–1900, 2000.

[8] H. A. Gieser, "ESD testing: HBM to very fast TLP", tutorial presented at the ISREF 2004.

[9] D. W. Greve, "Programming mechanism of polysilicon resistor fuses," *IEEE Trans. Electron Dev.*, vol. ED-29, No. 4, pp. 719–724, 1982.

[10] D. Pierce, "Electrostatic discharge (ESD) failure mechanisms," *EOS/ESD Symposium*, Tutorial Notes, pp. C-1–C-32, 1995.

[11] A. Wang, "On-chip ESD protection design: mixed-mode simulation and ESD for RF/mixed-signal ICs," tutorial presented at the BCTM 2003.

[12] C. Duvvury and A. Amerasekera, "ESD: a pervasive reliability concern for IC technologies," *Proc. of the IEEE*, vol. 81, No. 5, pp. 690–702, 1993.

[13] ITRS 2005, Factory Integration Chapter, Electrostatic Discharge – Backup Section: Details and Assumptions for Technology Requirements, http://www.itrs.net/Links/ 2005ITRS/Linked%20Files/2005Files/Factory/ElectroStatic%20Discharge%20Backup05RevFINAL.ppt

[14] W. Russell, "Defuse the threat of ESD damage," *Electronic Design*, vol. 43, No 5, pp. 115–120, 1995.

[15] M. Song, D. C. Eng, and K. P. MacWilliams, "Quantifying ESD/EOS latent damage and integrated circuit leakage currents," *EOS/ESD Symposium*, pp. 304–310, 1995.

[16] R. Renninger, M. Jon, D. Ling, et al. "A field induced charged-device model simulator," *EOS/ESD Symposium*, pp. 59–71, 1989.

[17] S. H. Voldman, "The impact of technology scaling on ESD robustness of aluminum and copper interconnects in advanced semiconductor technologies," *IEEE Trans. on Components, Packaging & Manufacturing Technology*, Part C (Manufacturing), vol. 21, No. 4, pp. 265–277, 1998.

[18] D. K. Kontos, R. Gauthier, D. E. Ioannou, T. Lee, Min Woo, K. Chatty, C. Putnam, and M. Muhammad, "Interaction between electrostatic discharge and electromigration on copper interconnects for advanced CMOS technologies," *Proc. of the Int. Reliability Physics Symp. (IRPS)*, pp. 91–97, 2005.

[19] C. -K. Hu, B. Luther, F. B. Kaufman, J. Hummel, C. Uzoh, and D. J. Pearson, "Copper interconnection integration and reliability," *Thin Solid Films*, vol. 262, No. 1–2, pp. 84–92, 1995.

[20] S. H. Voldman, "The impact of technology evolution and scaling on electrostatic discharge (ESD) protection in high-pin count high-performance microprocessors," *IEEE Int. Solid-State Cir. Conf. (ISSCC)*, pp. 366–367, 1999.

[21] S. H. Voldman, "A review of electrostatic discharge (ESD) in advanced semiconductor technology," *Microelectronics Reliability*, vol. 44, No. 1, pp. 33–46, 2004.

[22] Semiconductor Industry Association (SIA), "The National Technology Roadmap for Semiconductors," 2005.

[23] S. Voldman, "High-current characterization of dual-damascene copper interconnects in SiO2- and low-k interlevel dielectrics for advanced CMOS semiconductor technologies," *Proc. of the Int. Reliability Physics Symp. (IRPS)*, pp. 144–153, 1999.

[24] W. Fichtner, K. Esmark, and W. Stadler, "TCAD software for ESD on-chip protection design," *Proc. of the Int. Electron Devices Meeting (IEDM)*, pp. 14.1.1–14.1.4, 2001.

[25] H. Gossner, "ESD protection for the deep sub micron regime – a challenge for design methodology," *Proc. of the Int. Conf. on VLSI Design (VLSID)*, pp. 809–818, 2004.

Chapter 2

ESD MODELS AND TEST METHODS

1. INTRODUCTION

Electrostatic discharge (ESD) events are recognized as a significant contributor of early life failures and failures throughout the operating life of semiconductor devices. Although contemporary integrated circuit designs include ESD protection circuitry, the effectiveness of this protection must be determined in a manner which will ensure its effectiveness in the "real world" if the part is to meet the reliability requirements for the given application. An ESD event may carry amperes of current in a short period of time, typically from hundreds of pico-seconds to hundreds of nano-seconds. Needless to say such events are very harmful for sensitive electronic components and integrated circuits (ICs).

For the purposes of reproduction under controlled conditions, the real word ESD events are classifieds in three main categories under which electronic elements or ICs are tested:

- *The human body model* (HBM) represents an ESD event caused by a charged human discharging the current into a grounded IC.
- *The machine model* (MM) represents a discharge coming from a charged machine into a grounded IC. This ESD model is typically used in automotive assembly lines.
- *The charge device model* (CDM) covers the ESD discharge when a device or an IC is self-charged during the manufacturing process and comes into the contact with grounded equipment.

O. Semenov et al., ESD Protection Device and Circuit Design for Advanced CMOS Technologies, 21–43.

In the semiconductor industry, ±2 kV HBM, ±200 V MM and ±500 V CDM are very common requirements for general-purpose ICs. For some special applications, a much higher ESD protection level is required. For example certain automotive applications and smart card ICs require ±15 kV HBM ESD protection.

ESD qualification tests (HBM, MM and CDM) are Boolean, and often destructive in nature. User only gets the feedback whether or not the device under test (DUT) meets the ESD qualification criterion. Therefore, these tests are supplemented with non-destructive tests to collect additional information for analysis and design optimization. Obviously, detailed information on the ESD behavior of protection elements and circuits are required for their optimization. For such analysis and design optimization, Transmission line pulsing (TLP) technique is employed as an alternative and/or supplement to the model based ESD qualifications. The TLP technique has gained popularity in the semiconductor industry in recent years due to its flexibility and ease of generating pulses with different pulse widths and magnitudes. In addition, the TLP testing is not destructive in nature.

Finally, note that ESD tests are generally performed at the component level which includes discrete semiconductor devices and ICs; and at the printed circuit board (PCB) level. Recently, the charge board model (CBM) ESD testing was proposed for PCBs. This ESD model simulates the damage induced when objects at non-ground potential are almost instantaneously discharged. The industrial data show that board-mounted ICs that were robust to ESD at the component level could be damaged by ESD at the board level. The individual components mounted on the board have failures with a definite ESD-like signature, but they are more severe than the damages caused by HBM, MM, or CDM [1].

2. ESD ZAPPING MODES

The ESD voltage stress polarity could be positive or negative between two arbitrary pins of the chip. However, in general the device pins are stressed with respect to power supply or ground. Therefore, depending on the polarity of electrostatic charge and the discharge path, four possible zapping modes exist for an ESD event [2]. These modes are called PS-mode, NS-mode, PD-mode and ND-mode and are shown in Figure 2-1. The PS (NS) mode refers to the situation when a positive (negative) ESD voltage is applied to the DUT and the ESD current discharges through the V_{SS} pin: (i) the PS mode is the case when a positive ESD voltage is applied for a pin with the V_{SS} pin is grounded and the V_{DD} pin and other pins are floating; (ii) the NS mode describes the case when a negative ESD voltage applied for a pin with the V_{SS} pin is grounded, while the V_{DD} pin and other pins are floating.

This causes the ESD voltage stress to appear between V_{SS} and V_{DD} power lines. Figure 2-1(a) and (b) show these modes. Due to the parasitic resistance and capacitance along V_{SS}/V_{DD} power lines in CMOS ICs as well as the voltage drops on the input-to-V_{SS} and V_{DD}-to-V_{SS} ESD protection elements, such ESD discharging path had been reported to cause significant ESD damages on internal circuits beyond the input/output buffers and the ESD protection devices [3]. Similarly, the PD (ND) mode is a case, when a positive (negative) ESD voltage with respect to V_{DD} is applied to the DUT and the ESD current discharges through the grounded V_{DD} pin. PD and ND modes are shown in Figure 2-1(c) and (d), respectively.

The ESD protection circuits for advanced submicron CMOS ICs should perform the effective ESD discharging path from input and output pads to both V_{SS} and V_{DD} power lines. This is especially necessary for nanometer CMOS circuits with larger chip size and longer V_{DD} and V_{SS} power rails with higher parasitic resistances and capacitances [4].

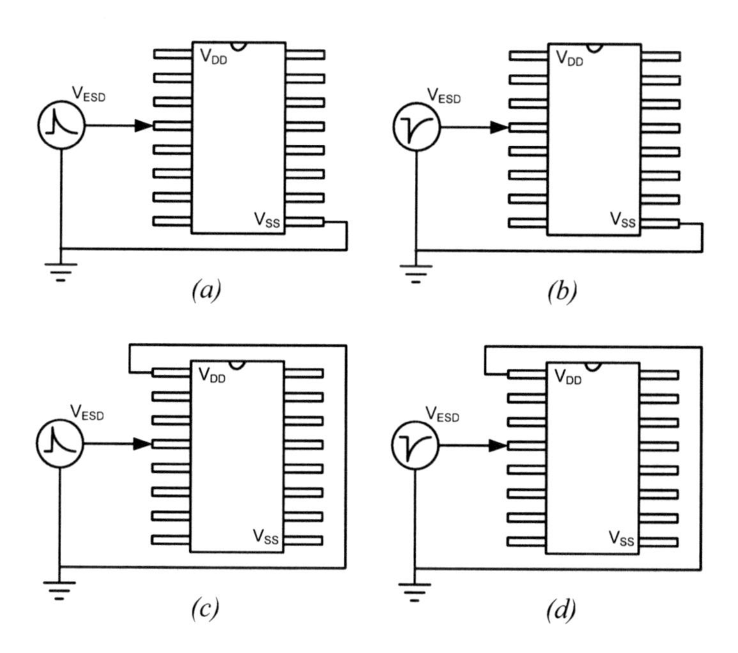

Figure 2-1. ESD zapping modes (a) PS-mode (b) NS-mode (c) PD-mode (d) ND-mode.

The ESD failure threshold of a pin is defined as the lowest (in absolute value) ESD-sustaining voltage of the four-mode ESD stresses on the pin. For example, if an output pin can sustain even up to 4 kV ESD voltage in the PS, NS, and PD mode ESD stresses, but can only sustain 1 kV ESD voltage in the ND mode ESD stress, the ESD failure threshold for this output pin is defined as 1 kV only.

3. HBM MODEL

Under various conditions, the human body can be charged with electrical energy and transfer that charge to a semiconductor device through normal handling or assembly operations. To evaluate the effectiveness of the protection circuitry in an integrated circuit, HBM ESD testing is performed. This HBM pulse is intended to simulate the human body type ESD conditions which the part would experience during normal usage. The ESD testing is also used to determine the immunity or susceptibility level of a system or part to the HBM ESD event. Several different Human Body Model (HBM) ESD simulation circuits and pulse waveforms exist, including Military Standard MIL-STD 883E (method 3015.7) [5], JEDEC STANDARD [6] and others.

Today, the HBM model is the most commonly used discharge model in the microelectronics industry. For the HBM ESD event, the test attempts to simulate what happens when a human becomes charged (through motion, walking, etc.), and then discharges by touching the conductive leads of a device. This device, at the time of the discharge, is assumed to be at a lower

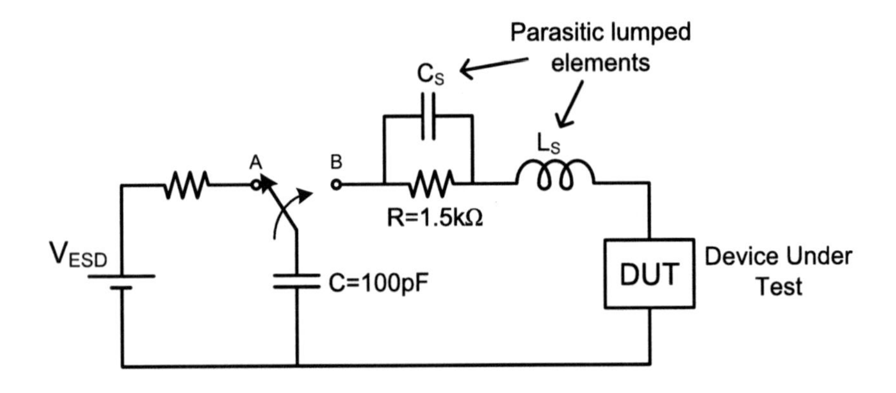

Figure 2-2. The HBM ESD equivalent circuit.

potential. The worst-case scenario is when the device is grounded. The equivalent HBM circuit includes the HBM resistor (R = 1,500 Ω), the HBM capacitor (C = 100 pF), the series inductor (L_s = 8 μH) and the optional stray capacitance of the HBM resistor (C_s = 1.5 pF) [7, 8]. The equivalent RLC circuit of HBM with parasitic lumped elements is shown in Figure 2-2.

Owing to the large serious resistance in the HBM, this ESD event can be modeled as a current source with the current waveform shown in Figure 2-3. The current peak is typically 1.20–1.48 A for 2 kV HBM ESD stress, the rise time is 2–10 ns and the decay time is 130–170 ns [6]. In real world situations, a person may have a resistance and capacitance that differ from this model. In addition, the ambient temperature and relative humidity also affect the current waveform. The amount of moisture in air controls the air resistance. At higher levels of humidity, it is more difficult to generate and sustain a charge. For example, walking across a carpet can generate 35 kV in low humidity compared to 1.5 kV in high humidity [9]. Also, the ESD current is increased with the increasing of ambient temperature. The temperature increase at constant humidity leads to an increase in the kinetic energy of ions in air and hence higher velocity streamers. The higher ions velocity results in the decreasing of air resistance [10]. In order to minimize errors with environmental conditions, all testing must be performed in a controlled environment room which maintained temperature and humidity at relatively constant levels of 23 ± 4°C and 32 ± 5% relative humidity [11]. As a result, the true human body ESD waveform can be significantly different from that specified in HBM ESD standards.

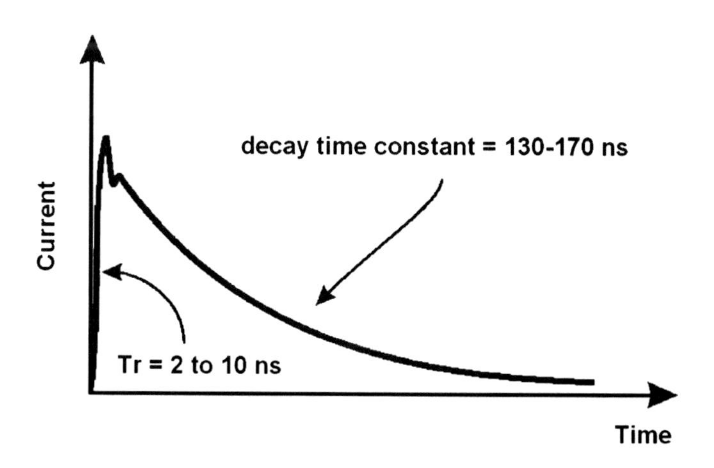

Figure 2-3. Current waveform of HBM ESD event.

HBM testing is performed in accordance with the ESD association specification [12]:

- Each I/O pin should be stressed against each power supply pin. It is permissible to short all the power supply pins together.
- Each power supply should be stressed with respect to other power supplies.
- Three repeated ESD zaps in sequence are required, and there should be at least a 300 ms interval between consecutive zaps.

Based on the ESD protection level, HBM model has different classes. Table 2-1 shows these different classes along their protection level.

Table 2-1. Human body model classes.

Class	**Voltage range (V)**
Class 0	<250
Class 1A	250–500
Class 1B	500–1,000
Class 1C	1,000–2,000
Class 2	2,000–4,000
Class 3A	4,000–8,000
Class 3B	>8,000

This classification allows easy grouping and comparing of components according to their ESD protection level.

4. MM MODEL

The machine model is similar to the human body model, with the substitution of a 1,500 Ω resistor by the 750 nH inductor and increasing the capacitance to 200 pF, as it is shown in Figure 2-4. This model is intended to represent the type of damage caused by equipment used in manufacturing. As apparent from the model, higher capacitance and lower overall impedance of the path results in higher current densities during the MM discharge. Therefore, even though MM damages are similar to that of HBM, they occur at much lower threshold levels. Typical voltage values on MM model range from 100 to 500 V. The inductance value in the model is the most critical parameter because it controls the rise time of the current waveform during the discharge. Because of the very small series resistance in the MM, this ESD event can be modeled as an ideal voltage source. The comparison of HBM and MM current waveforms are depicted in

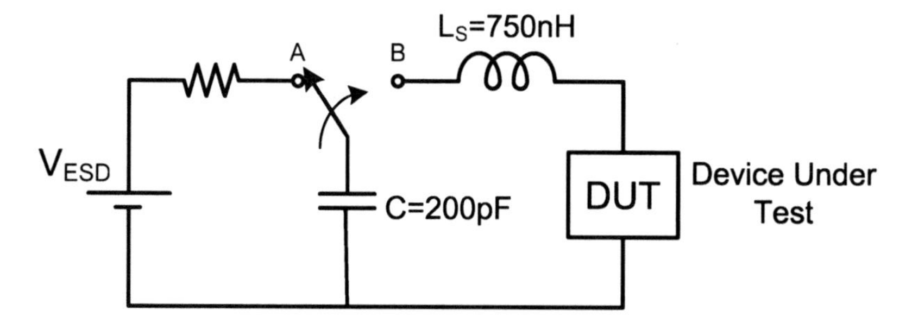

Figure 2-4. The MM ESD equivalent circuit.

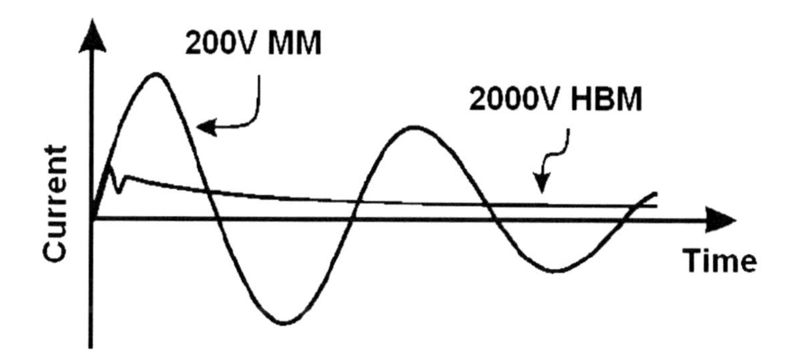

Figure 2-5. Current waveform of MM ESD event.

Table 2-2. Machine model classes.

Class	Voltage range (V)
Class M1	<100
Class M2	100–200
Class M3	200–400
Class M4	>400

Figure 2-5. MM waveforms are defined by JEDEC [13], IEC [14] and ESDA [15] ESD standards. The main problem with MM is that the parasitic resistance of real testers is not zero. Together with the fact that MM addressed the same failure modes in the device as HBM [16], testing with the HBM seems the preferable method for characterizing the ESD sensitivity of ICs and PCBs.

Similar to HBM model, MM model is also classified based on the ESD protection level. Table 2-2 shows these classes.

5. CDM MODEL

The CDM is the newest model and its ESD stress is the most difficult to reproduce through a tester. It is very sensitive to parasitics in the test hardware. This model is used to simulate the event that occurs from charged packaged ICs which subsequently discharging into a low impedance ground. This ground can be a hard grounded surface or a large charge sinks like a metal work table or tool. This could occur during testing where a part is charged during the assembling process or on the shipping tubes [17, 18]. When the part comes into contact with the tester's pins, it discharges. The discharge impedance in real life is close to zero, but it is finite and small in an ESD tester. This is where the variability between testers comes into play. The simple equivalent circuit of the CDM ESD is depicted in Figure 2-6. In this figure, C_{CDM} is the sum of all capacitances in the device and the package with respect to ground and R_{CDM} is the total resistance of discharge path. For 500 V CDM, the typical model parameters are C_{CDM} = 10 pF, R_{CDM} = 10 Ω, R_L = 10 Ω, and L_S = 10 nH. The current waveform of a 500 V ESD event has a rise time of ~0.3 ns and a peak of ~10.4 A.

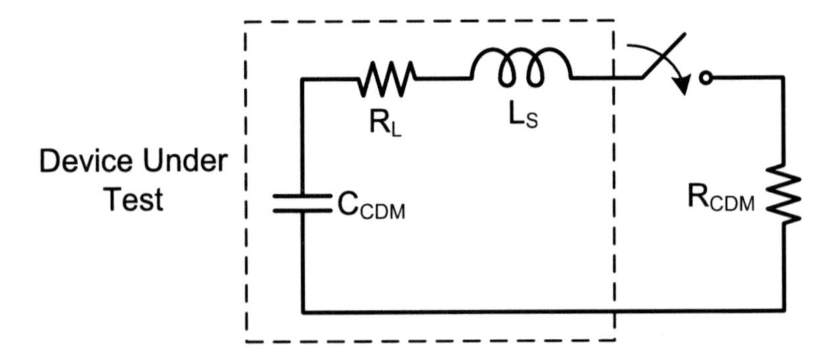

Figure 2-6. The CDM ESD equivalent circuit.

In a CDM ESD stress, the peak current is much higher compared to the current in an HBM stress. Moreover, the rise time as well as the stress duration is much shorter, as it is shown in Figure 2-7. The rise time is limited by the inductance and resistance of the current path. For many packages, the bond wire inductance of ~0.5 nH that limits the rise and fall times. With progress of automation in manufacturing lines, the HBM and MM damages are expected to decrease and major failures are likely to be caused by the CDM. Widely used automatic manufacturing and test equipment generate static charges and can damage the integrated circuits [19]. Once the ICs become charged, they can rapidly discharge as they come

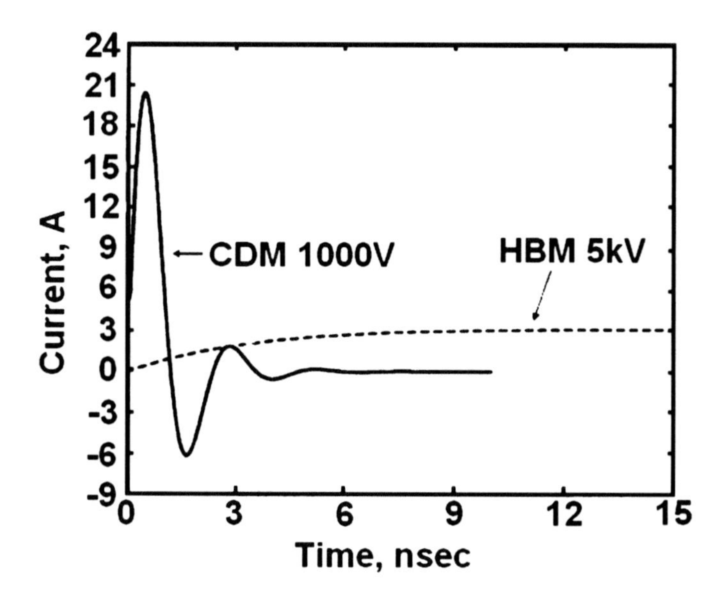

Figure 2-7. Current waveform of CDM ESD event. (Adapted from [18].)

Table 2-3. Charged device model classes.

Class	Voltage range (V)
Class C1	<125
Class C2	125–250
Class C3	250–500
Class C4	500–1,000
Class C5	1,000–1,500
Class C6	1,500–2,000
Class C7	>2,000

into contact with a grounded test head or metal surface of Surface Mount Technology (SMT) machine, resulting in a CDM ESD event. Furthermore, as the dielectric failure has gained in scaled semiconductor devices [20], the CDM accounts for the majority of ESD failures during chip manufacturing.

CDM model is classified based on the ESD protection level as well. Table 2-3 shows these classes. The protection level of 500 V for CDM ESD stress is typically used to ensure the general-purpose chip reliability. Higher CDM classes are applied for special ICs used for automotive, aviation and other industries with high reliability requirements.

There are two methods for CDM ESD testing that depend on how the part is held during the ESD testing. They are referred to as socketed and non-socketed methods [21]. As it is suggested by their names, in non-socketed method the test is applied to the device directly while in socketed method the test is applied to the device through the socket. Therefore, non-socketed method is a better model for real world CDM event. In this method the test is done by placing the device upside down (pins facing up) on a field plate and charging and discharging it as shown in Figure 2-8 [22].

In the socketed method, similar to HBM and MM test, the device is placed on a socket and the whole board is charged and discharged during the test. Therefore, the discharge current is generated not only from the device but also from the parasitic capacitance and inductance of the socket and the simulator network. The socketed CDM test produces more severe damage and in some cases a different failure mode compared to the non-socket test [21, 23]. As this method is not a real CDM test, it is usually referred to as Sockted Device Model (SDM).

Currently, there are several CDM standards developed by JEDEC solid state technology association [24] and ESD association [25]. However, the existing documents are still debated between the physical background of the event and the freedom of designing and building a test system that can be used in industrial environment for CDM testing.

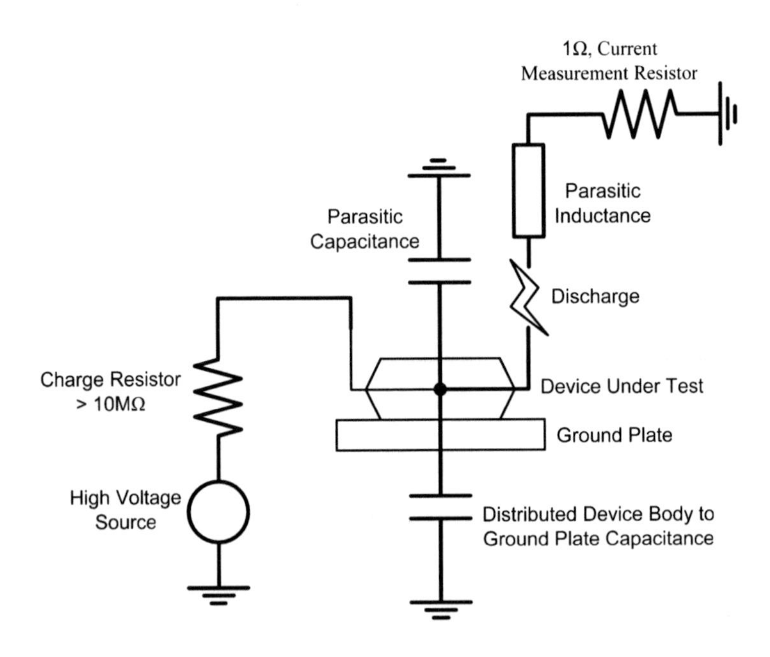

Figure 2-8. Typical non-socketed CDM test setup.

6. CBM MODEL

Recent industrial data indicates that HBM rarely mimics real word ESD failures [18]. HBM ESD damage is no longer the most prevalent one for numerous reasons. Most component manufacturers and users have effective controls against HBM ESD events, and the last generations of packages with mm-range dimensions are often too small for people to handle with fingers. Even in cases of relatively large components, most high volume component and board manufacturing uses automated equipment so humans rarely touch components. Most real word component ESD failures can be simulated by the Charged Device Model (CDM) or by the Charged Board Model (CBM) if components mounted on a circuit board. It was shown that ICs robust to HBM and CDM damage at the package level may be susceptible to CBM damage at the board level depending on the PCB design or board capacitance [1, 26]. Testing to the CBM event is similar to testing to the CDM event. Since there were no standards to follow, CBM stress-testing methods and procedures have been developed by in-house design and product groups. No specific, standardized procedures yet exist for testing these board related ESD damages. One of the first field-induced charged board model (FCBM) simulator was developed in AT&T Bell Laboratories [26] (see Figure 2-9). The CBM simulates a charged PCB discharging just before contact is made with a conductive object at or near ground potential.

A CBM discharge has much higher energy than a CDM discharge for a given charge voltage, since PCB capacitance is much higher than IC package capacitance. In addition, CBM discharges generally have faster

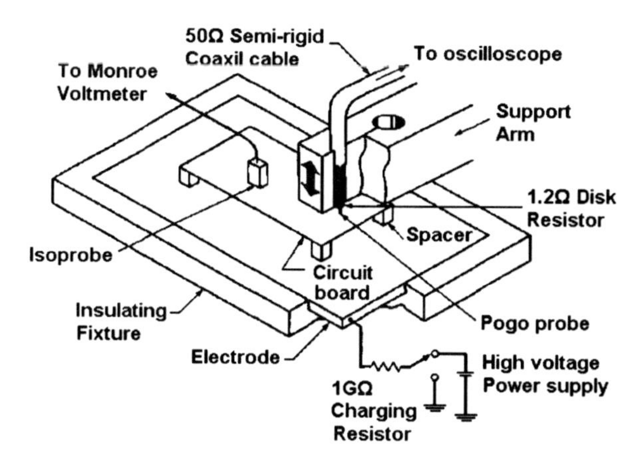

Figure 2-9. Field-induced CBM simulator. (Adapted from [26].)

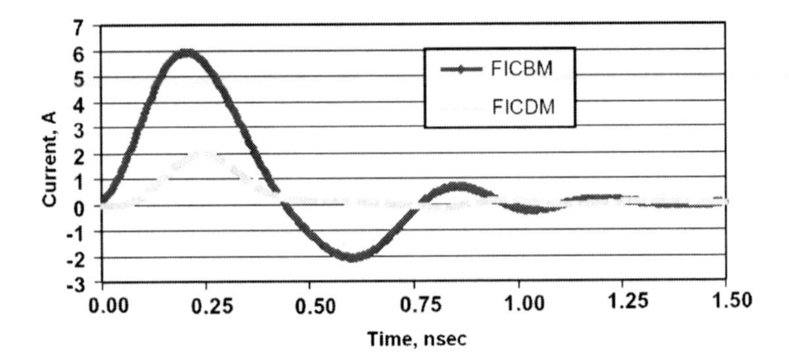

Figure 2-10. Comparison of FICBM and FICDM current discharge waveforms of Dual Op Amp with a 375 V charge voltage. (Adapted from [1].)

rise times than CDM discharges. Thus, ICs that are effectively immune to ESD damage at the device-level may be susceptible to ESD damage at the board-level. For example, the comparison of CBM and CDM discharge waveform of Dual Op Amp is shown in Figure 2-10. In this figure, the CBM current waveforms of the Dual Op Amp assembled in a 8-lead SOIC package and the Dual Op Amp packaged and soldered at the center of a small (3 in. × 3 in. × 60 mil) two-layer FR4 board are compared [1]. The difference in the discharge waveform peak currents is due to the larger effective capacitances by ~3.8× at the board level in comparison to the chip level. Due to the high energy associated with real world PCB discharges, the CBM ESD damage can be more severe than typical device-level ESD damage. Consequently, such ESD damage can be easily mistaken for EOS damage.

Finally, it was found that the CBM susceptibility depends on the overall area and layout of the PCB power planes. Larger power planes typically result in larger capacitance, thus resulting in lower overall IC CBM withstand voltages than smaller power planes.

7. TLP TESTING

The integrated circuit industry has been using transmission-line pulse (TLP) testing to characterize on-chip electrostatic discharge (ESD) protection structures since 1985. This TLP ESD testing technique was introduced by Maloney and Khurana as a new electrical analysis tool to test the many single elements used as ESD protection structures [27]. IC designers now use TLP testing in increasing numbers, because it provides a reliable, repeatable, and constant amplitude waveform. The TLP testing lets ESD engineers more accurately measure the conditions that cause IC failures. TLP uses the rectangular-pulse testing (RPT) or square-pulse testing (SPT) to simulate the energy in an exponential HBM test pulse. There is a correspondence between TLP (with rectangular pulse widths of 75–200 ns) and HBM (with a 150 ns exponential pulse width). The ESD industry has settled on a 100 ns pulse width and 2–10 ns rise time TLP pulses as the de facto standard [28]. Correlation is further achieved by comparing the rise times of both systems, because the rise time of either threat pulse can cause significant differences in device failures. The typical TLP and HBM current waveforms are shown in Figure 2-11. The correlation issues of different ESD models and TLP testing are discussed in details in the next section. Note, that the TLP waveform doesn't represent any real world ESD event. It's strictly a tool that designers of ESD protection structures can use to perform circuit analysis and failure analysis. Recently, the ESD Association published the release of the ESD Association's Transmission Line Pulse (TLP) ESD Standard Practice document, which can be used as a first industrial standard for TLP testing [29].

To produce a TLP waveform, a TLP tester charges a transmission line with a high-voltage DC source. When the transmission line discharges, the pulse it creates injects current into the DUT. The original TLP tester designed by Intel's engineers injected a constant current into the DUT for the length of the pulse [27]. Figure 2-12 shows a schematic of a constant-current TLP tester. A 50 Ω resistor provides a known, controlled load for the transmission line. A resistor in series with the DUT converts the voltage

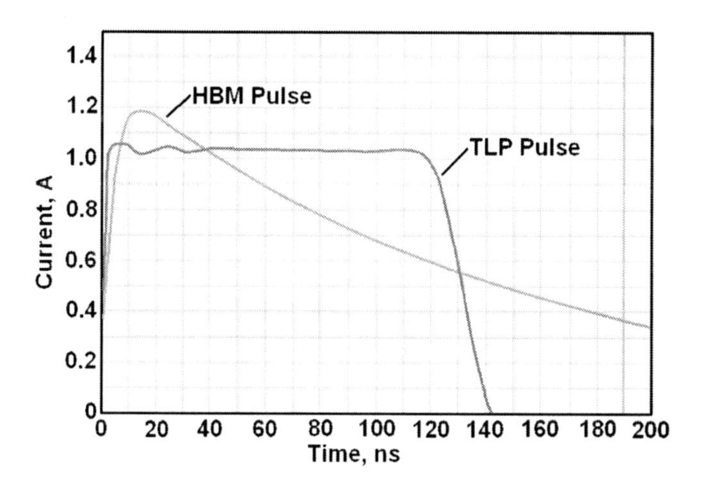

Figure 2-11. HBM pulse vs. TLP pulse. (Adapted from [28].)

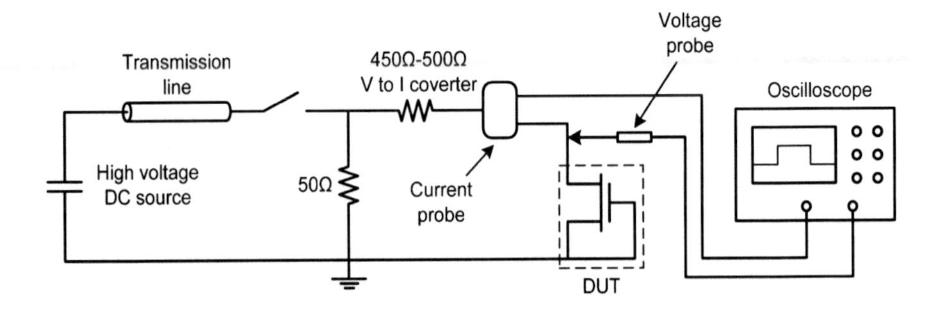

Figure 2-12. Schematic of a constant-current TLP tester. (Adapted from [28].)

from the transmission line into a somewhat constant current before injecting it into the DUT. A constant-current TLP tester can deliver up to about 4 A to the DUT, but many IC designs require higher current. To obtain higher current, some engineers use constant-impedance TLP testers. The constant-impedance TLP tester can provide up to 10 A. Low-pass filters alter the pulse's rise time, typically over a range of 0.2–10 ns. The technical details of currently available TLP testers are explains in recently published reviews [28, 30].

A TLP tester must build an I-V curve through numerous sets of measurements. This curve represents accurate measurements of both the voltage across the device and the current through the device by averaging the digitized data for some length of time near the end of the pulse (pick-off points range from 50% to 90% of the pulse width). For each point on the curve, the tester applies a TLP pulse to the ESD protection structure under test while the device is non-powered. TLP can measure and display the dc leakage current of the DUT at nominal power supply. This leakage current measurement is performed after each TLP pulse. Adding a dc leakage current data provide additional insight into minute changes in damage to the protection or core circuits, which is unavailable without this measurement. The dc leakage current data combined with the I-V data provide electrical indications of where damage begins, and how rapidly it can evolve from soft to hard failure [31]. These I-V and leakage data on a device are defined as the electrical damage signature, as it shown in Figure 2-13. In this figure, V_{t1} is the triggering voltage of ESD device, V_h is the holding voltage, and I_{t2} is the second (thermal) breakdown current. Breakdown occurs when the leakage current suddenly increases, which usually occurs at point I_{t2} (the second snapback point) of the I-V curve of ESD device.

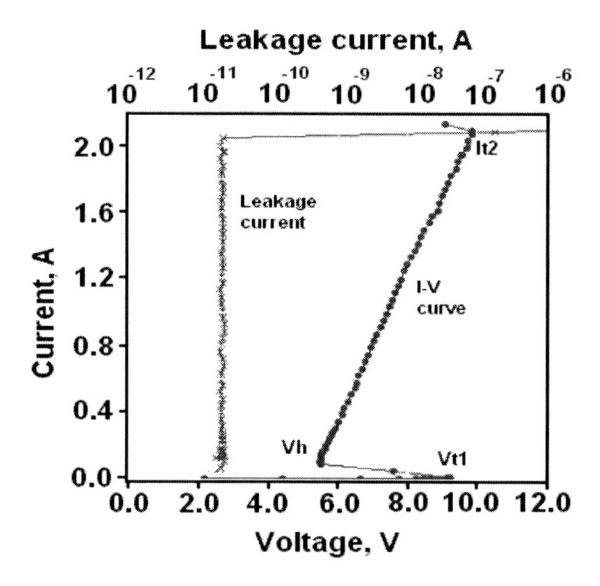

Figure 2-13. Typical TLP breakdown I-V characteristics and leakage current curve.

Researchers have attempted to correlate the TLP measurement results with CDM testing as well. Given the fast rise and fall times, the TLP techniques should give us a better understanding of the dynamic effects such as trigger speeds and transient currents under ESD CDM type conditions. The trigger speed of a device can be very well analyzed from the TLP Voltage–Time response. However, there is the fundamental difference between CDM and TLP. CDM is a one pin test, while TLP requires the selection of two IC pins. Therefore, the ESD discharge path for both CDM and TLP are different within an IC. Nevertheless, very fast TLP offers opportunities to the ESD engineer to deduce CDM relevant parameters of the ESD protection design, such us the transient voltage response. To characterize devices in the time and current domain of CDM, a measurement method is necessary that applies equivalent conditions to the device under test. It is also important that the measurement delivers data with a high enough resolution to define the transient behavior of that device. A technique that complies with those requirements is a very fast TLP (vf-TLP) [32, 33]. In this system, a square pulse of short duration and rapid rise time, typically it's used a pulse width of 10 ns with a rise time of 750 ps, is applied to the device under test by discharging a pre-charged transmission line into the load device. This incident pulse is reflected at the device under test. Unlike conventional TLP with 100 ns pulse width the DUT does not often reach an equilibrium state during a vf-TLP pulse. Typically, the average current and voltage over the time interval between 65% and 95% of the pulse widths is used to determine a single IV data point at the vf-TLP IV characteristics [33].

The TLP test method is the only non-destructive technique to study the internal ESD behavior in integrated circuits. TLP can be regarded as an engineering and design tool, whereas the HBM stress test is a qualification tool that provides levels of threshold failures (that is, classes such as 1, 2, 3, etc.). TLP testing provides detailed knowledge about the actual ESD current paths, which is not available from traditional ESD tests on ICs. This knowledge is essential for understanding the behavior of both passing and failing products. Note, that the TLP characterization of standalone protection devices and complete circuits can be different. This is because the chip level TLP measurements include the additional information on the actual operation of the ESD protection devices placed in ICs, such as parallel current paths, rail bus resistances, different behavior of identical pads located in different segments of I/O ring and other issues [34]. The gained insight gives valuable suggestions to designers and product engineers for improving circuit designs and protection strategies.

8. CORRELATION OF ESD TEST METHODS

As it was previously mentioned different methods for testing of ESD robustness and sensitivity are in common use today, including Human Body Model (HBM), Machine Model (MM), Charged Device Model (CBM) and Transmission Line Pulse testing (TLP). Many researchers have studied the correlation of the various methods. It was shown that in order to make a valid comparison of the test results it is required that the current path and the failure mechanism are identified unambiguously [35]. The special test structures should be used to ensure that there is only one possible current path in the protection device during ESD event. And the failure mechanism should be identified by combining electrical measurements with thorough physical failures of the damage site. Once these conditions are met, a correlation analysis between the different ESD test methods can be done. In this section we consider ESD test methods, which are relatively well correlated to each other, such as HBM-MM, HBM-TLP and CDM-vf-TLP. Other combinations of test methods have relatively poor correlations.

8.1 HBM and MM Correlation

M. Kelly et al. analyzed six different devices implemented in different CMOS technologies [36]. These devices also had different package types. The failure threshold voltage levels for these devices were obtained by HBM and MM testing. These threshold voltage levels were used to perform the correlation analysis. Failure mode analysis procedure was performed to establish a physical signature associated with electrical failure of devices at the ESD threshold. It was found that the failure signatures show moderate overlap between the HBM and MM ESD stressing. It should be noted that even a complete overlap may not guarantee a good correlation. Using the failure threshold values, a correlation coefficient analysis was done and a high correlation between HBM and MM ESD testing was found. The correlation coefficient was 0.93 and a regression model was $V_{HBM} = (11.64 \div 11.95) \times V_{MM}$. The high correlation coefficient (0.87) for HBM and MM testing was also obtained by G. Notermans et al. [35].

8.2 HBM and TLP Correlation

Typically, TLP systems have 100 ns wide rectangular pulses because this length pulse has been found to initiate junction damage at the same peak current as HBM test pulses [37]. The 100 ns wide rectangular TLP pulse has been shown to provide correlation to the HBM pulse. This conclusion was done based on TLP and HBM testing of different devices implemented in 0.35 μm [37] and 0.18 μm [38] CMOS technologies. However, the TLP pulse width and also the IC technology may have an impact on the correlation factor. Generally, the physical correlation between HBM and TLP can be established by the comparison of the physical failure signature after failure analysis. For 0.5 and 0.18 μm CMOS technologies, the correlation model was $V_{HBM}(kV) = 2.1 \times I_{TLP}(A)$ and $V_{HBM}(kV) = (1.53 \div 1.56) \times I_{TLP}(A)$, respectively. The correlation coefficients for these technologies were found as 0.96 and $1.43 \div 1.71$, respectively [35, 38].

Recently, the relationship between HBM and TLP testing of low temperature poly-Si thin film transistors (LTPS-TFT) was investigated [39]. It was found that the peak current in TFT device is increased linearly with the increase of HBM ESD zapping voltage. Hence, the device turn-on resistance (R_{HBM}) under HBM ESD stress can be calculated by the following equation: $R_{HBM} \approx (1/I_{ESD}) \times (V_{ESD} - I_{ESD} \times 1.5\ k\Omega)$. In case of TLP testing, the device turn-on resistance can be extracted from the slope of TLP measured I-V curves. The ratio of R_{TLP} and R_{HBM} of analyzed TFTs was found approximately 1.2. Hence, the HBM and TLP tests of LTPS-TFTs had almost perfect agreement.

8.3 CDM and vf-TLP Correlation

The CDM stress has a high peak current up to 10 A which is reached within a range of a nanosecond. Due to the completely different current and time domain, the CDM testing does not correlate well with HBM, MM and conventional 100 ns TLP testing. For example, the correlation coefficients between CDM-HBM, CDM-MM, are 0.28 and 0.42, respectively [36]. To characterize devices in the time and current domain of CDM, a measurement method is necessary that applies equivalent conditions to the device under test. The vf-TLP technique satisfies these requirements. The effectiveness of this technique was demonstrated on a relatively old 1 μm CMOS process. It was found that the failure voltages of CDM and SDM are between 6 and 8.5 times higher than the vf-TLP (3.5 ns pulses) failure voltages. The failure current threshold of vf-TLP is from 1.3 to 1.7 times higher than the CDM failure current [32]. Generally, the correlation of vf-TLP testing with the CDM/SDM testing is achieved in terms of the failure signature. The failure thresholds of vf-TLP and CDM/SDM do not correlate due to the different current paths [32, 33]. In the CDM, charges are injected or coming out of one pin and the full device is maintained at some fixed potential. In the vf-TLP, charges flow between the stress pin and a reference pin for the duration of the square pulse.

9. ESD TESTERS

In the previous sections different methods to evaluate ESD robustness have been discussed. There are a wide range of testers in industry that have been designed to target a specific model in ESD area. As an ESD protection designer, HBM, MM, CDM and TLP test are the most important evaluation methods for ESD robustness. There are three major classes of tester available in industry to address these tests: HBM/MM testers, CDM testers and TLP testers.

- *HBM/MM testers*: This family of testers is usually checking the device under test (DUT) for HBM or MM model. The results are usually as a pass or fail for a particular voltage level. Generally, they provide up to 8 kV HBM voltage test. Oryx instruments and Thermo electron corp. are among the leading manufacturers of HBM/MM testers. Figure 2-14 shows an ICMS-700 HBM/MM tester produced by Oryx Instruments Corporation [40].
- *CDM testers*: Similar to HBM/MM testers, CDM testers also provide a pass or fail result for different voltage levels. They normally allow a test with a 50 V to 2 kV CDM stress. Figure 2-15 shows an Orion CDM tester developed by Oryx instruments.

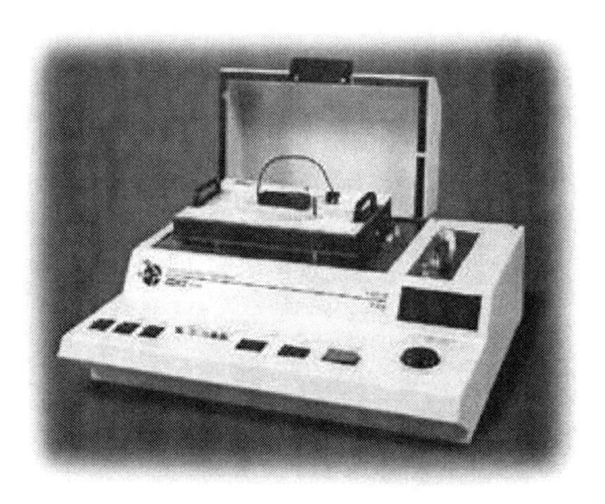

Figure 2-14. ICMS-700 HBM/MM tester.

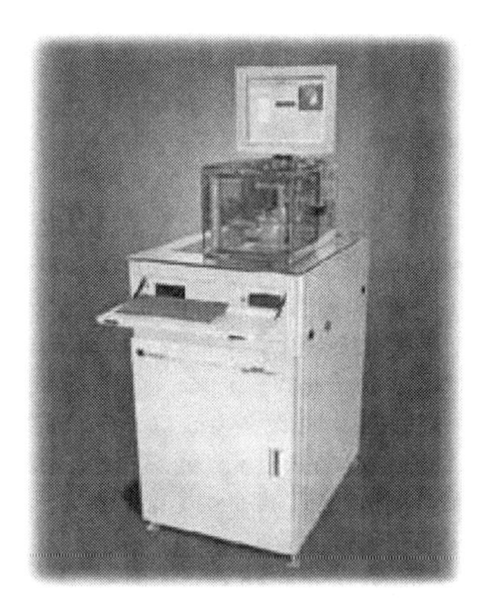

Figure 2-15. Orion CDM tester.

- *TLP testers*: Unlike HBM/MM and CDM testers, TLP testers show the complete characteristics of ESD protection circuits. Therefore, in order to measure the I-V characteristic of ESD protection circuits TLP testers should be used. Barth Electronics Inc. [41], SQP Products [42] and Oryx Instruments are the main manufacturers of commercial TLP testers. Figure 2-16 shows a Barth 4002 TLP tester by Barth Electronics Inc.

Figure 2-16. Barth 4002 TLP tester.

10. SUMMARY

Currently, three different ESD models (HBM, MM, and CDM/CBM) are used to emulate the real word ESD events. The IC industry has standardized these models to define how charge is transferred during an ESD event. The charge transferring in these models can be modeled, in a first order of approximation, by an RLC network. However, these developed models do not cover all possible variables that influence a real ESD event. As such, results obtained using these test methods should be used for comparisons of the ESD robustness of various designs and not as an absolute measure of a parts capability in the real word environment [43].

The square pulsing ESD test techniques, such as TLP and vf-TLP, are the standard analysis tools for characterizing the protection device behavior at the ESD event, particularly turn-on, high current and leakage current characteristics. In the other words, the TLP test method is the only technique to study the internal ESD behavior in integrated circuits. TLP testing provides detailed knowledge about the actual ESD current paths, which is not available from traditional ESD tests on ICs.

Many researchers have studied the correlation of the various ESD test methods. It was shown that in order to make a valid comparison of the test results it is required that the current path and the failure mechanism are identified unambiguously. The analysis of experimental data shown that HBM and MM, HBM and TLP and CDM and vf-TLP test methods have relatively good correlated each to other. Other combinations of test methods have a poor correlation. The following regression models and conclusions were obtained:

- $V_{HBM} = (11.64 \div 11.95) \times V_{MM.}$
- $V_{HBM}(kV) = 2.1 \times I_{TLP}(A)$ or $V_{HBM}(kV) = (1.53 \div 1.56) \times I_{TLP}(A)$ for 0.5 μm and 0.18 μm CMOS technologies, respectively.
- The correlation of vf-TLP testing with the CDM/SDM testing is achieved in terms of the failure signature. The failure thresholds of vf-TLP and CDM/SDM do not correlate.

REFERENCES

[1] A. Olney, B. Gifford, J. Guravage, and A. Righter, "Real-word printed circuit board ESD failures," *Microelectronics Reliability*, vol. 45, No. 2, pp. 287–295, 2005.

[2] M. -D. Ker, H. -H. Chang, and C. -Y. Wu, "A gate-coupled PTLSCR/NTLSCR ESD protection circuits for deep-submicron low-voltage CMOS IC's," *IEEE J. of Solid-State Cir.*, vol. 32, No. 1, pp. 38–51, 1997.

[3] C. C. Johnson, T. J. Maloney, and S. Qawami, "Two unusual HBM ESD failure mechanisms on a mature CMOS process," *EOS/ESD Symposium*, pp. 225–231, 1993.

[4] H. Terletzki, W. Nikutta, and W. Reczek, "Influence of the series resistance of on-chip power supply busses on internal device failure after ESD stress," *IEEE Trans. Electron Dev.*, vol. 40, No. 11, pp. 2081–2083, 1993.

[5] Military Standard MIL-STD 883E (method 3015.7), 1996.

[6] JEDEC STANDARD, JESD22-A114D (Revision of JESD22-A114-C.01), 2006.

[7] L. van Roozendaal, A. Amerasekera, P. Bos, W. Baelde, F. Bontekoe, P. Kersten, E. E. Korma, P. Rommers, P. Krys, U. Weber, and P. Ashby, "Standard ESD testing of integrated circuits," *EOS/ESD Symposium*, pp. 119–130, 1990.

[8] K. Verhaege, P. J. Roussel, G. Groeseneker, H. E. Maes, H. Gieser, C. Russ, P. Egger, X. Guggenmos, and F. G. Kuper, "Analysis of HBM ESD testers and specifications using a 4th order lumped element model," *EOS/ESD Symposium*, pp. 129–137, 1993.

[9] R. Zezulka, "ESD basics," *1993 EOS/ESD Symposium*, Tutorial Notes, pp. A-1–A-28, 1993.

[10] I. A. Metwally, "Factors affecting corona on twin-point gaps under dc and ac HV," *IEEE Trans. on Dielectrics and Electrical Insulation*, vol. 3, No. 4, pp. 544–553, 1996.

[11] M. A. Kelly, G. E. Servais, and T. V. Pfaffenbach, "An investigation of human body electrostatic discharge," *Int. Symp. for Testing and Failure Analysis (ISTFA)*, pp. 167–173, 1993.

[12] ESD sensitivity testing: Human Body Model (HBM) – Component level, ESD Association standard, S5.1, 1993.

[13] JEDEC STANDARD, Electrostatic discharge sensitivity testing machine model, JESD22-A114-A, 1998.

[14] IEC, Electrostatics, Part 3.1, Methods for simulating of electrostatic effects – Machine model, Component testing, 61340-3-2, 2002.

[15] ESD Association, ESD standard test method for electrostatic discharge sensitivity testing – machine model, ESD STM5.2, 1998.

[16] M. Beh, C. Kang, M. Natarajan, and M. K. Radhakrishnan, "Analysis of HBM and MM ESD failures in nMOS devices," *Int. Symp. on the Physical and Failure Analysis of Integrated Circuits (IPFA)*, pp. 111–115, 1995.

[17] R. G. Renninger, M. C. Jon, D. L. Lin, T. Diep, and T. L. Welsher, "A field-induced charged-device model simulator," *EOS/ESD Symposium*, pp. 59–71, 1989.

[18] J. Lee, K. -W. Kim, Y. Huh, P. Bendix, and S. -M. Kang, "Chip-level charged-device modeling and simulation in CMOS integrated circuits," *IEEE Trans. on Computer-Aided Design of Integrated Circuits and Systems*, vol. 22, No. 1, pp. 67–81, 2003.

[19] J. Bernier and B. Hesher, "ESD improvements for familiar automated handlers," *EOS/ESD Symposium*, pp. 110–117, 1995.

[20] C. Leroux, P. Andreucci, and G. Reimbold, "Analysis of oxide breakdown mechanism occurring during ESD pulse," *IEEE Int. Reliability Physics Symp.*, pp. 276–282. 2000.

[21] A. Olney, "A combined socketed and nonsocketed CDM test approach for eliminating real-world CDM failures," *EOS/ESD Symposium*, pp. 62–75, 1996.

[22] ESD Association, http://www.esda.org/documents/esdfunds5print.pdf

[23] K. Verhaege, C. Russ, J. -M. Luchies, and G. Groeseneken, "Grounded-Gate nMOS Transistor behavior under CDM ESD stress conditions," *IEEE Trans. on Electron Dev.*, vol. 44, No. 11, pp. 1972–1980, 1997.

[24] JEDEC STANDARD, "JESD22-C101-A: Field-induced charged device model test method for electrostatic discharge-withstand thresholds of microelectronic components," 2000.

[25] ESD Association, "ESD STM5.3.1-1999: Standard test method for electrostatic discharge sensitivity testing – charged device model (CDM) component level," 1999.

[26] D. L. Lin, "FCBM – a field-induced charged-board model for electrostatic discharges," *IEEE Trans. on Industry Applications*, vol. 29, No. 6, pp. 1047–1052, 1993.

[27] T. Maloney and N. Khurana, "Transmission line pulsing techniques for circuit modeling of ESD phenomena," *EOS/ESD Symposium*, pp. 49–54, 1985.

[28] M. Rowe, "TLP testing gains momentum," Test & Measurement World, Sept. 2002.

[29] ESD Association, "ESD SP5.5-TLP: Standard Practice for Electrostatic Discharge (ESD) Sensitivity Testing, Transmission Line Pulse (TLP) Testing – Component Level," 2002.

[30] L. G. Henry, J. Barth, K. Verhaege, and J. Richner, "Transmission-Line Pulse ESD Testing of ICs: A New Beginning," *Compliance Engineering*, 2001.

[31] L. G. Henry, J. Barth, J. Richner, and K. Verhaege, "Transmission line pulse testing of the ESD protection structures of ICs—a failure analyst's perspective," *Int. Symp. for Testing and Failure Analysis (ISTFA)*, pp. 203–213, 2000.

[32] H. Gieser and M. Haunschild, "Very fast transmission line pulsing of integrated structures and the charged device model," *IEEE Trans. on Components, Packaging, and Manufacturing Technology*, Part C, vol. 21, No. 4, pp. 278–285, 1998.

[33] J. Willemen, A. Andreini, V. De Heyn, K. Esmark, M. Etherton, H. Gieser, G. Groeseneken, S. Mettler, E. Morena, N. Qu, W. Soppa, W. Stadler, R. Stella, W. Wilkening, H. Wolf, and L. Zullino, "Characterization and modeling of transient device behavior under CDM ESD stress," *Journal of Electrostatics*, vol. 62, No. 2–3, pp. 133–153, 2004.

[34] T. Smedes, R. M. D. A. Velghe, R. S. Ruth, and A. J. Huitsing, "The application of transmission line pulse testing for the ESD analysis of integrated circuits," *Journal of Electrostatics*, vol. 56, No. 3, pp. 399–414, 2002.

[35] G. Notermans, P. de Jong, and F. Kuper, "Pitfalls when correlating TLP, HBM and MM testing," *EOS/ESD Symposium*, pp. 170–176, 1998.

[36] M. Kelly, G. Servais, T. Diep, D. Lin, S. Twerefour, and G. Shah, "A comparison of electrostatic discharge models and failure signatures for CMOS integrated circuit devices," *Journal of Electrostatics*, vol. 38, No. 1–2, pp. 53–71, 1996.

[37] A. Amerasekera, L. van Roozendaal, J. Abderhalden, J. Bruines, and L. Sevat, "An analysis of low voltage ESD damage in advanced CMOS processes," *EOS/ESD Symposium*, pp. 143–150, 1990.

[38] J. Barth, K. Verhaege, L. G. Henry, and J. Richner, "TLP calibration, correlation, standards, and new techniques," *EOS/ESD Symposium*, pp. 85–96, 2000.

[39] M. -D. Ker, C. -L. Hou, C. -Y. Chang, and F. -T. Chu, "Correlation between transmission line pulsing I-V curve and human body model ESD level on low temperature poly-Si TFT devices," *Int. Symp. on the Physical and Failure Analysis of Integrated Circuits (IPFA)*, pp. 209–212, 2004.

[40] http://www.oryxinstruments.com

[41] http://www.barthelectronics.com

[42] http://www.sqpproducts.com

[43] K. Verhaege, C. Russ, D. Robinson-Hahn, M. Farris, J. Scanlon, D. Lin, J. Veltri, and G. Groeseneken, "Recommendations to furter improvements of HBM ESD component level test specifications," *EOS/ESD Symposium*, pp. 40–53, 1996.

Chapter 3

ESD DEVICES FOR INPUT/OUTPUT PROTECTION

1. INTRODUCTION

As discussed in Chapter 2, ESD is a very high current event. Therefore, ESD protection circuits should be able to handle a large amount of current without being destroyed. A number of semiconductor devices can be used to safely sink (source) this current; hence can be used as ESD protection circuit. In this chapter, some of the most important devices that are used in CMOS ESD protection circuits are discussed. Unlike conventional MOS transistors, the ability to carry large current is the most important design attribute of these devises.

Based on the shape of the *I-V* characteristic of semiconductor devices, they are divided into two main categories: non-snapback devices and snap-back devices.

2. NON-SNAPBACK DEVICES

In this category of devices if the voltage across them is increased beyond a certain voltage, the current starts to increase rapidly while the voltage across them remains constant. Diodes are good examples of this category. As an ESD protection device, *p-n* junction diode, zener diode, and punch through TVS are widely used and are explained in this section.

O. Semenov et al., ESD Protection Device and Circuit Design for Advanced CMOS Technologies, 45–83.

2.1 P-N Junction Diode

A *p-n* junction diode is a simple and effective ESD protection device. Their forward bias behavior is exploited to safely carry a large amount of ESD current while their reverse bias behavior is exploited under normal operating conditions. In this section, these two operating regions are discussed and compared under high current conditions.

2.1.1 Forward-Biased Diode

The biasing and the *I-V* characteristic of a forward-biased diode are shown in Figure 3-1. The forward-biased diode can conduct significant current when the applied voltage is greater than V_{on} which is normally in the 0.5–0.7 V range. At this point, the resistance of the diode is very low and usually less than 1 Ω. In order to formulate the on-resistance and on-voltage of a forward biased diode consider an abrupt junction diode with the well knows current equation [1]:

$$I_D = I_S\left(e^{V_D/V_t} - 1\right)$$

$$I_S = qA\left[\frac{D_p p_{no}}{L_p} + \frac{D_n n_{po}}{L_n}\right] = qAn_i^2\left[\frac{1}{N_D}\sqrt{\frac{D_p}{\tau_p}} + \frac{1}{N_A}\sqrt{\frac{D_n}{\tau_n}}\right],$$

where V_t is the thermal voltage and is 26 mV at room temperature, D_p and D_n are diffusion coefficients of holes and electrons respectively, and τ_p and τ_n are hole and electron lifetime. The value of on-voltage and on-resistance of this diode can be calculated as:

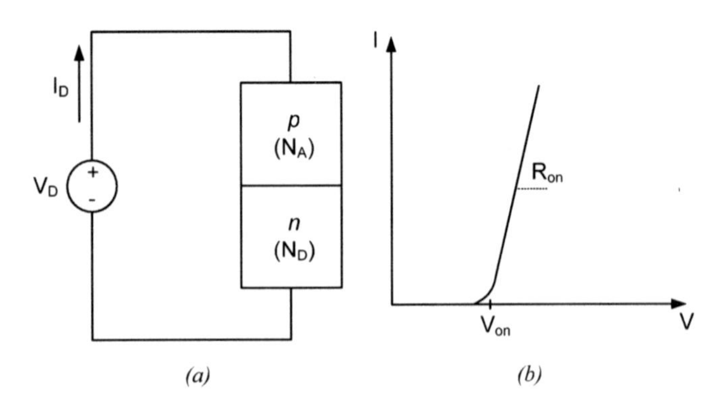

Figure 3-1. Forward-biased p-n junction diode (a) schematic (b) I-V characteristic.

$$V_{on} = V_t \ln \frac{N_A N_D}{n_i^2} \tag{3-1}$$

$$R_{on} = \left[{dI_D}/{dV_D} \right]^{-1} = \frac{V_t}{I_S} e^{-V_D/V_t}, \quad V_D > V_{ON}$$

It can be seen that the on-voltage and the on-resistance are functions of the semiconductor doping and junction area and, therefore, process dependent. As an ESD protection device, a diode should be able to carry a large current in forward bias. Due to low on resistance and low on-voltage of the forward-biased diode, power dissipation and, therefore, internal temperature of the diode remains low under high current conditions. Hence, the maximum current carrying capability of the forward-biased diode is high and usually in the order of 20–50 mA/μm [2]. Thus, the forward biased diode is a very promising device for ESD protection application.

2.1.2 Reverse-Biased Diode

Bias conditions and the *I-V* characteristic of a reverse-biased diode are shown in Figure 3-2. In this operating region, unlike the forward-biased diode, current is generated through the avalanche breakdown of the *p-n* junction. In this device the on-voltage equals to the avalanche breakdown voltage of the *p-n* junction which is calculated from equation (3-2) for an abrupt junction diode [38].

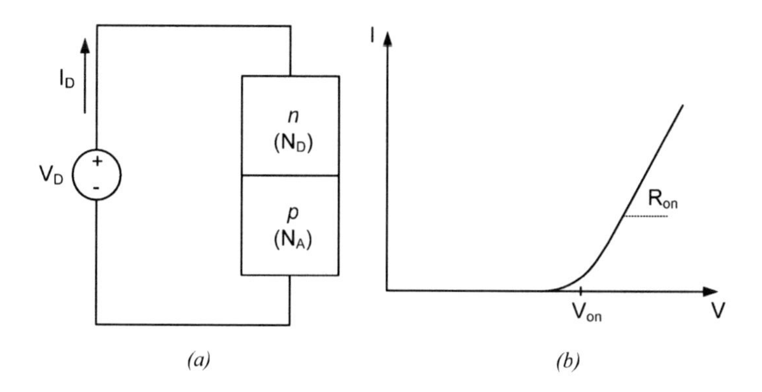

Figure 3-2. Reverse-biased p-n junction diode (a) schematic (b) I-V characteristic.

$$BV = \frac{\varepsilon(N_A + N_D)}{2qN_A N_D} E_{crit}^2 = 3.25 \times 10^6 \frac{N_A + N_D}{N_A N_D} E_{crit}^2 \quad (3\text{-}2)$$

Typical values for the on-voltage and the on-resistance for this diode is in the range of 10–20 V, and 50–100 Ω respectively. Similar to the forward-biased diode, these values are functions of the semiconductor doping as well as the junction area.

Unlike the forward-biased diode, the reverse-biased diode has high on-resistance and high on-voltage. Therefore, power dissipation and internal temperature of the reverse-biased diode are high under high current conditions. Hence, the maximum current carrying capability of this diode is very low and in the order of 0.5–2 mA/μm, making it unsuitable for ESD protection applications [2]. On the other hand this diode shows small current when used in normal circuit operation, i.e. the voltage across it is V_{DD}. Therefore, a *p-n* junction diode can be used in such a way that under ESD conditions it is biased in forward biased region while under normal operating conditions it is biased in reverse biased region. Further discussion on this application is presented in Chapter 5.

2.1.3 Diode in Standard CMOS Technology

In a standard single well CMOS process with a p-type substrate, three different diffusion diodes can be created: n^+-diode, p^+-diode, and n-well diode. Figure 3-3 shows the cross section of these diodes.

The n^+-diode is formed between an n^+ junction and the p-substrate as shown in Figure 3-3(a). As substrate should be connected to ground/V_{SS} in CMOS technology this diode can be used only between the pad and

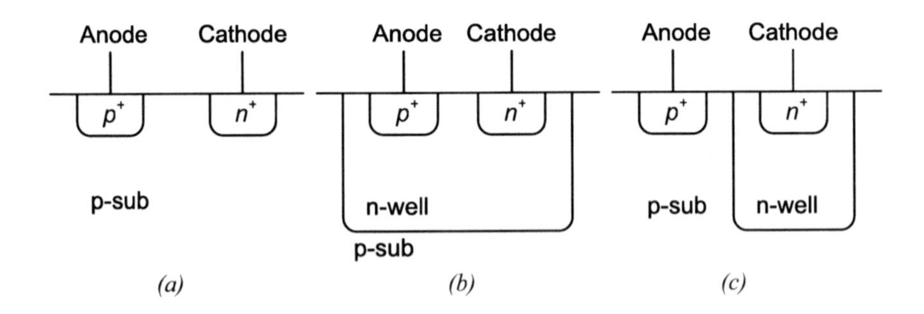

Figure 3-3. Cross section of diodes in CMOS technology: (a) n^+-diode (b) p^+-diode (c) n-well diode.

ground/V_{SS}. As this diode has a small junction area it cannot carry high amount of current. Figure 3-3(b) shows the p^+-diode which is formed between a p^+ and an n-well region. Again as n-well region should be connected to V_{DD} cathode of this diode should be connected to V_{DD}. Hence this diode can be used between the pad and V_{DD}. Finally, Figure 3-3(c) shows the cross section of an n-well diode which is formed between an n-well and the p-sub regions.

These diodes can be compared based on their parasitic capacitance and their ESD protection level. The p^+-diode and the n-well diode have higher ESD protection levels due to their larger junction area. On the other hand, this larger junction area causes higher parasitic capacitance. In order to do a comprehensive comparison, a new figure of merit, which is defined as the ratio of HBM protection level over parasitic capacitance, has been introduced [3]:

$$\text{FOM1'} = \frac{V_{HBM}}{C_{ESD}}$$

The n^+-diode and p^+-diode have the highest value of FOM1'. In 0.18 μm CMOS technology, FOM1' for the n^+-diode and p^+-diode is reported as 43.9 and 30.6 V/f F, respectively [3].

2.2 Zener Diode

In addition to *p-n* junction diode, zener diode can also be used in ESD protection circuits. Zener diode is a reverse biased diode with lower triggering voltage. In conventional reverse biased *p-n* junction, the diode is formed between a high-doped (p^+ or n^+) and a low-doped region (n-well or p-sub). Based on equation (3-2), in order to reduce the breakdown voltage of the reverse biased diode both *n* and *p* regions should be highly doped. In a standard CMOS technology there are two methods to realize a *p-n* junction between two high doped regions:

The first method uses a Lightly Doped Drain (LDD) region to create the zener diode [4]. LDD region is a low doped extension of n^+ or p^+ regions. The main purpose of this region is to reduce hot carrier effect [5]. Figure 3-4 shows the cross section of this diode.

It can be seen that silicide block option has been used in this diode. Silicidation has become a standard CMOS process step in submicron technologies. Silicidation is the addition of a Tungsten or Cobalt interface to the semiconductor material. It is usually applied to the polysilicon gate and

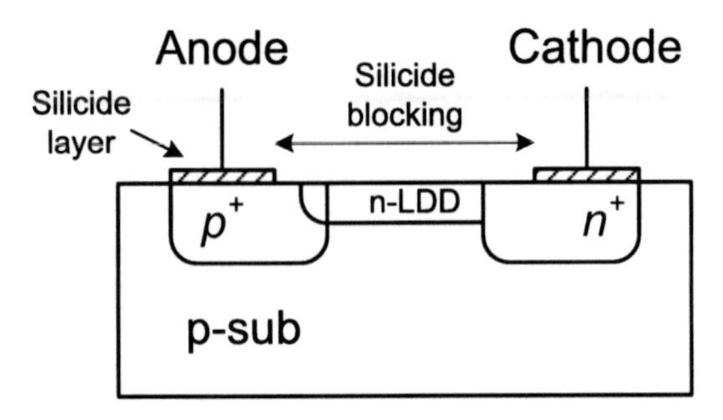

Figure 3-4. Zener diode with LDD.

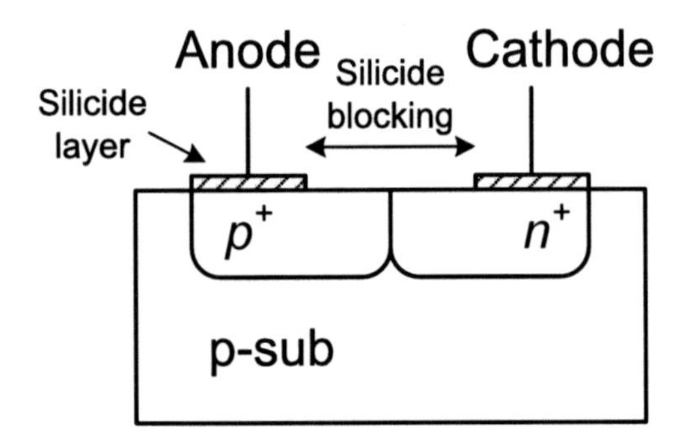

Figure 3-5. Zener diode.

diffusions. The silicided poly and diffusions have a sheet resistance of at least one order of magnitude lower than non-silicided ones which improves transistor switching speed. Referring to the cross section of this zener diode, it can be seen that there is an overlap between *p* and *n* diffusion regions. Therefore, in a silicided process, *p* and *n* regions will be shorted through the silicide layer. As a result, silicide block option should be used as shown in Figure 3-4. Silicide block is not a part of standard CMOS technology and requires one extra mask which increases the fabrication cost.

In the second method, the zener diode is realized by placing an n^+ diffusion region next a p^+ diffusion region [6]. Figure 3-5 shows the cross section of this diode. Similar to the zener diode shown in Figure 3-4, this diode needs silicide block option as well.

Although the zener diode has lower on-voltage compared to regular reverse-biased *p-n* junction diode, its on-voltage is still higher than oxide breakdown voltage and hence, it is not used as the main protection device.

This diode is usually used as a secondary device which helps the main protection device.

2.3 Polysilicon Diode

The polysilicon diode is another p^+-n^+ diode which is created in the polysilicon layer instead of the silicon substrate [7]. As the doping of these n^+ and p^+ regions is the same as that of polysilicon gates of NMOS/PMOS transistors, this diode is fully compatible with the standard CMOS process. Figure 3-6 shows the cross section of this diode.

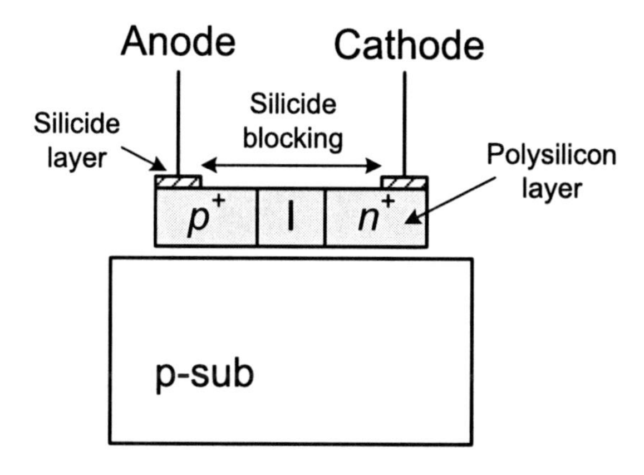

Figure 3-6. Cross section of the polysilicon diode.

As the polysilicon region is silicided in submicron CMOS technologies, silicide block option is required to avoid shorting the anode and the cathode of the diode. It can be seen that an un-doped region called 'I' exists between the n^+ and p^+ regions. The triggering voltage of this diode is a function of the length of the 'I' region. When the n^+ and p^+ regions have direct connection the width of I is zero and when there is an overlap between n^+ and p^+ regions the value of I is negative. Figure 3-7 shows the triggering voltage of the forward biased polysilicon diode for different 'I' values in 0.25 μm CMOS technology [7].

It can be seen that as the spacing between n^+ and p^+ junctions is increased, the forward biased breakdown voltage is increased as well. The reason for this increase is that the undoped region acts as a barrier for carriers to cross the intersection. Hence, the spacing 'I' can be adjusted to obtain the required breakdown voltage. The main advantage of this diode over the regular *p-n* junction diode is its better substrate noise coupling

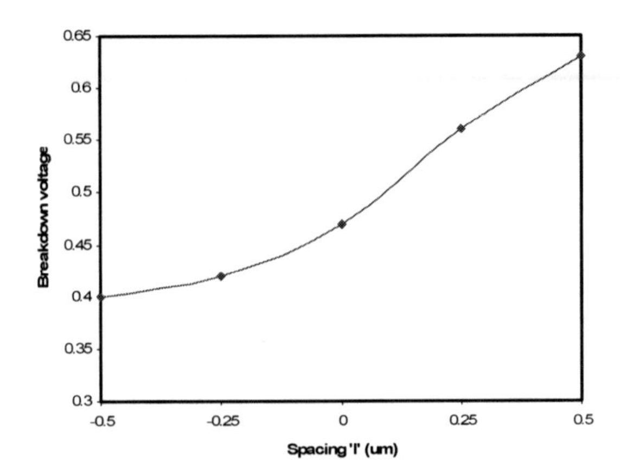

Figure 3-7. Impact of n^+*-*p^+ *spacing on breakdown voltage.*

immunity and lower leakage current. The reason is that the polysilicon diode has no junctions in the substrate and is isolated from the substrate.

2.4 Stacked Diodes

In high frequency applications, instead of using one forward biased diode, two or more diodes are used in series [8]. This technique can be applied to either regular *p-n* junction or polysilicon diodes. Figure 3-8 shows a stack of *n* diodes along with their parasitic capacitance and resistance.

For the stacked diode, considering all diodes are the same, the overall parasitic resistance and capacitance can be calculated from the following equations.

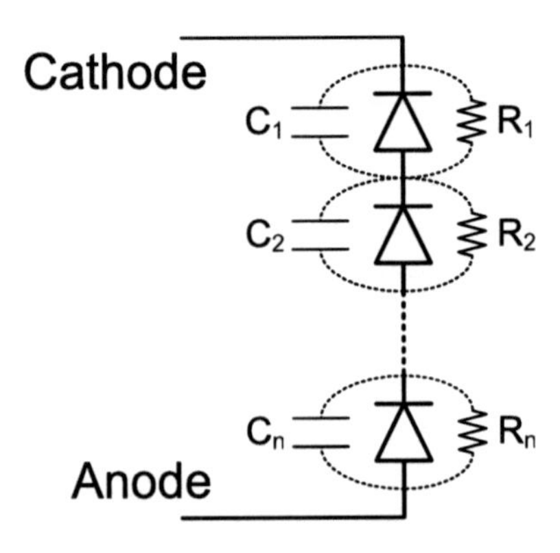

Figure 3-8. Stacked diodes.

$$R_p = R_1 + R_2 + \ldots + R_n = nR_1 \tag{3-3}$$

$$C_p = \frac{1}{1/C_1 + 1/C_2 + \ldots + 1/C_n} = \frac{C_1}{n} \tag{3-4}$$

It can be seen that by stacking n diodes, the parasitic capacitance is divided by n while the parasitic resistance is multiplied by n. Measurement results show that the reduction trend of parasitic capacitance is saturated as the number of diodes increases to more than three [8]. This is due to the presence of parasitic capacitors other than junction capacitors which are not in series. Having lower parasitic capacitance makes this configuration suitable for high frequency applications. On the other hand, higher parasitic resistance and larger area are the major limitations of this method. Furthermore, measurement results show a slight decrease in I_{t2} by increasing the number of diodes. This reduction is 1% for three diodes which can be neglected [8]. Therefore, there is a trade off between the number of stacked diodes and secondary effects. It's been shown that a stack of three diodes gives the optimum results [8].

3. SNAPBACK DEVICES

Snapback devices, similar to a reverse biased diode, go to the breakdown region as the voltage across them is increased. After breakdown, due to an internal positive feedback mechanism, the voltage across the device drops and the device moves from breakdown region to the holding region. MOSFET and Silicon Controlled Rectifier (SCR) are the most important snapback devices in the CMOS technology.

3.1 MOSFET

The simplest form of a MOSFET in ESD protection applications is the grounded gate configuration where the gate and source of the transistor are connected to ground. Figure 3-9(a) shows the cross section of a grounded gate NMOS (GGNMOS). The DC characteristic of the GGNMOS is shown in Figure 3-9(b).

In order to understand the behavior of the GGNMOS, consider Figure 3-9(a) where the parasitic bipolar transistor of the NMOS structure is explicitly shown. As the drain voltage increases, the drain-substrate junction becomes more reverse biased until it goes into avalanche breakdown. At this

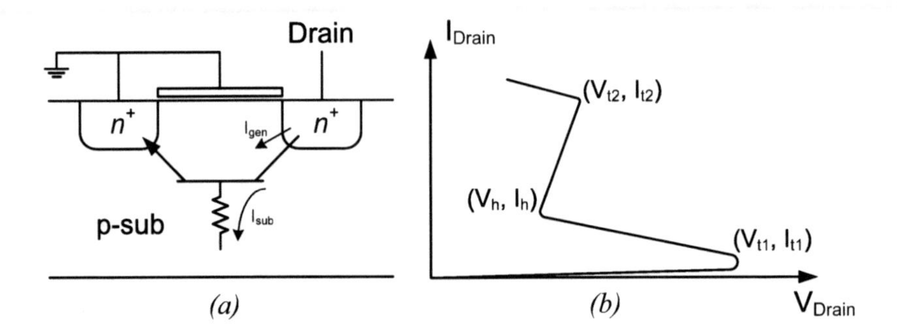

Figure 3-9. Grounded-gate NMOS (a) cross section (b) I-V characteristic.

point, the drain current increases, and the generated holes drift towards the substrate contact (I_{sub}), thereby increasing the base voltage of the parasitic bipolar transistor, which makes the base-emitter junction of the parasitic bipolar transistor more forward biased. As the base-emitter voltage reaches ≈0.7 V, the parasitic bipolar transistor turns on. The drain voltage at this point is V_{t1}. This bipolar action generates more current, and therefore, there is no need to keep the drain voltage at V_{t1} to maintain the drain current. Hence, the drain voltage reduces to V_h, and snapback behavior will be observed. When the bipolar transistor turns on, increasing the drain voltage further increases the current until thermal damage occurs. This point is called the second breakdown point and the voltage and current at this point are V_{t2} and I_{t2}, respectively.

A basic analysis on the breakdown of a MOSFET can give a better insight of the breakdown mechanism [9]. In order to analyze this behavior, Figure 3-9(a) is redrawn in a schematic form in Figure 3-10.

In this figure R_{sub} is the substrate resistance and the current source I_h represents the hole current created by impact ionization near the drain junction. Using this diagram the body-source voltage of the NMOS transistor (base-emitter of the parasitic bipolar transistor) can be written in the following form:

$$V_{bs} = I_{sub} R_{sub} - V_{SB} \tag{3-5}$$

where V_{SB} is the external source to bulk bias.

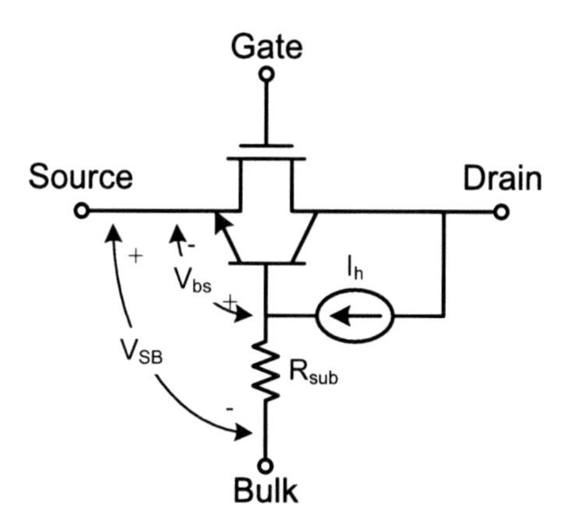

Figure 3-10. Equivalent circuit for an NMOS.

On the other hand bulk substrate current is the difference between the hole current created by impact ionization (I_h) and the base current of the parasitic bipolar transistor, which in turn consists of the current from base recombination and the current due to hole injection into the source. This leads to

$$I_{sub} = I_h - (1 - \gamma\alpha_T)I_e \tag{3-6}$$

where γ is the injection efficiency of the source junction, α_T is the base transport factor and I_e is the total current injected across the source junction. I_e is simply source-substrate junction current which is in the following form

$$I_e = I_0(\exp(V_{bs}/V_t) - 1) \tag{3-7}$$

In this equation I_0 is the reverse saturation current of the source-substrate junction and V_t is the thermal voltage.

The value of the current source can written in the following form:

$$I_h = (M - 1)(I_D + k\gamma\alpha_T I_e) \tag{3-8}$$

where M is the avalanche multiplication factor and k represents the ratio of electrons collected by the drain that go through the high-field region to the total number of electrons collected by the drain. By knowing the I_h, I_e and substrate current the total drain current can be written as follows

$$I_{DT} = M(I_D + k\gamma\alpha_T I_e) + \gamma\alpha_T(1 - k)I_e \tag{3-9}$$

In the next step, the conditions for breakdown can be studied. If the body-source voltage (base-emitter of the bipolar transistor) is less than 0.6 V, bipolar transistor is OFF and I_e is negligible. Therefore, V_{bs} increases linearly with M-1 and the currents would be

$$I_{sub} = (M-1)I_D$$

$$I_{DT} = MI_D$$

As V_{bs} reaches the turn-on voltage of the bipolar transistor (approximately 0.6 V) the bipolar transistor turns on and V_{bs} will be fixed at approximately 0.65 V. By substituting equations (3-3), (3-4) and (3-5) into (3-2) and knowing that $V_{bs} = 0.65$ V the electron current can written as

$$I_e = \frac{(M-1)I_D R_{sub} - V_{SB} - 0.65}{R_{sub}\left[1-\gamma\alpha_T - (M-1)k\gamma\alpha_T\right]} \quad (3\text{-}10)$$

Breakdown occurs when the electron current becomes very large. In other words, by setting the denominator of (3-7) to zero the breakdown condition can be calculated as:

$$M-1 = {(1-\gamma\alpha_T)}/{k\gamma\alpha_T} = 1/{k\beta}$$

where β is the effective common emitter current gain of the bipolar transistor when the MOSFET is operating in the saturation region.

From the above discussions, the conditions for breakdown can be summarized as follows:

1. The body-source (base-emitter of the bipolar transistor) junction must be turned on ($V_{bs} > 0.65$).
2. The multiplication factor must be significantly large to generate the required positive feedback.

For heavily doped substrates R_{sub} and thus the base voltage is small and the first factor is the dominant requirement for breakdown. But for low-doped substrates, where R_{sub} is large, and for high gate voltages, where I_D is large the second factor is the dominant requirement for breakdown.

As an ESD protection device, the drain is connected to the I/O pad. Therefore, under normal operating conditions, the NMOS transistor is OFF and the current of the ESD protection device is very small. Under ESD conditions and when the pad voltage exceeds V_{t1}, the transistor will go into snapback mode, and ESD current will be discharged through GGNMOS.

The maximum ESD current that can be discharged through this transistor is determined by the value of the second breakdown current. This value is usually in the order of 3–10 mA/μm. The width of the GGNMOS can be calculated based on the required ESD protection level. For example, to achieve 2 kV HBM protection and considering $I_{t2} = 4$ mA/μm, the width of the transistor can be calculated from the following equations:

$$V_{HBM} = \left(R_{ON(ESD)} + R_{HBM}\right) I_{t2}$$

$$R_{ON(ESD)} << R_{HBM} = 1.5\ \text{k}\Omega \Rightarrow W = 333\ \mu\text{m}$$

As it can be seen from the above example, the required width of a GGNMOS is typically a few hundred microns. Therefore, GGNMOS is usually realized in a multi-finger configuration.

From the DC characteristic of the GGNMOS shown in Figure 3-9(b), V_{t1}, V_h, V_{t2}, and I_{t2} are the most important parameters. In order to protect a circuit against ESD, the following requirements must be met.

1. V_{t1} must be less than the gate oxide breakdown voltage to protect the gate during an ESD event. This assures that the GGNMOS will turn on before the gate oxide breaks down.
2. V_{t2} must be greater than V_{t1} to ensure uniform triggering. This ensures that even if one finger triggers first, the voltage build-up can turn on other fingers before the first finger reaches the second breakdown region. Otherwise, the effective width of the device is decreased and the performance of the GGNMOS will be degraded.
3. I_{t2} determines robustness of the ESD protection device and should be as high as possible.
4. V_h should be greater than V_{DD}. Otherwise, GGNMOS may turn on during normal operating conditions and lead to latch-up. Typically, V_h should be designed to be 10–20% more than V_{DD}.

Practically, it is difficult to implement a GGNMOS in deep submicron CMOS technologies which meets all the above requirements. Hence, some new techniques have been developed to modify the GGNMOS structure in order to meet all the requirements for ESD protection.

3.1.1 GGNMOS in Advanced CMOS Technologies

In advanced CMOS technologies a number of new process steps have been added to the standard CMOS technology to increase the performance of the transistors. At the same time, some of these process steps have negative impact on ESD protection devices. Silicidation is the most critical process step that affects ESD protection devices. As mentioned earlier the main purpose of Silicidation is to increase the speed of integrated circuits by reducing the sheet resistance of poly and diffusion. However, as an ESD protection device, the resistance between gate and drain contacts, which is called the ballast resistance, forces a more uniform current flow through the NMOS transistor. As a result, all fingers of the transistor will trigger uniformmly. In non-silicided technologies, this resistance is created by the spacing between the gate and drain contacts. In silicided technologies, due to the silicidation of the diffusion regions, the resistance created by this spacing is very small and cannot ensure uniform triggering. There are a number of solutions for this problem.

1. N-well resistor: In this method, a ballast resistor is added externally. Figure 3-11 shows the layout of an NMOS with the n-well ballast resistor [11].
2. Silicide blocking: As mentioned in Section 2, using silicide block option, which requires one extra mask and three process steps (deposition of nitride, photo mask and etching), the ballast resistance can be restored [10].

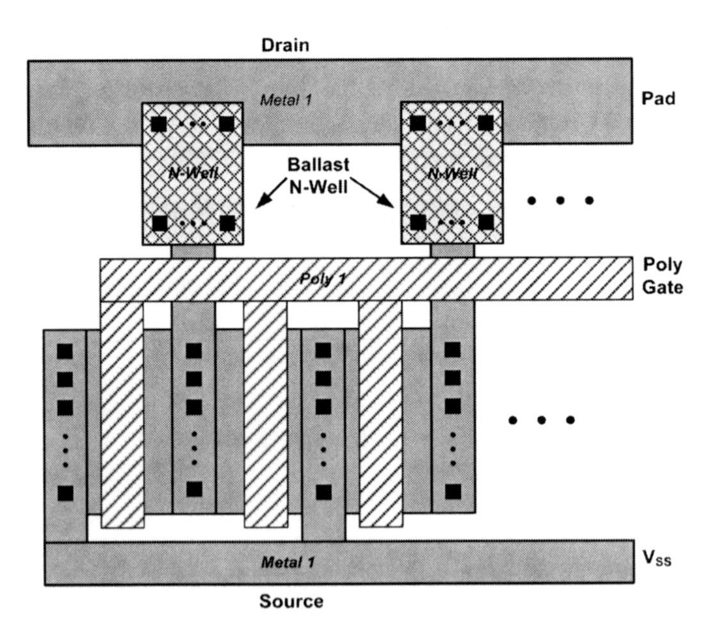

Figure 3-11. n-well resistor to create ballast resistor.

3. Backend ballasting: This method uses the high resistivity of contacts, vias and interconnects in advanced CMOS technologies to build the ballast resistor. Therefore, a chain of metal-inter-connect layers is used in this method [12].

3.2 Silicon Controlled Rectifier

Silicon Controlled Rectifier (SCR) is another active device that is often used as a protection element. Figure 3-12(a) shows a cross section of a simple SCR in standard CMOS technology. It consists of a *pnpn* structure. The p^+ diffusion in the n-well forms the anode, and the n^+ diffusion in the p-sub forms the cathode of the SCR. As an ESD protection device, the n-well contact is connected to the anode, and the p-sub contact is connected to the cathode. The anode is connected to the I/O pad, and the cathode is connected to the ground. SCR is often represented with its parasitic bipolar transistors as shown in Figure 3-12(b).

As the anode voltage increases, the n-well to p-sub junction becomes increasingly more reverse biased until it goes into avalanche breakdown. The generated current can turn on either of the two parasitic bipolar transistors. Typically, the gain of the *npn* transistor is an order of magnitude higher than that of the *pnp* transistor. Therefore, the *npn* transistor turns on easily compared to the *pnp* transistor. When the *npn* transistor turns on, its current generates a voltage drop across $R_{n\text{-well}}$ and turns on the *pnp* transistor as well. The current of the *p-n-p* transistor, in turn, creates a voltage drop across $R_{p\text{-sub}}$ and helps to keep the *n-p-n* transistor on. At this point, due to the current of the *pnp* transistor, there is no need for the anode to provide the

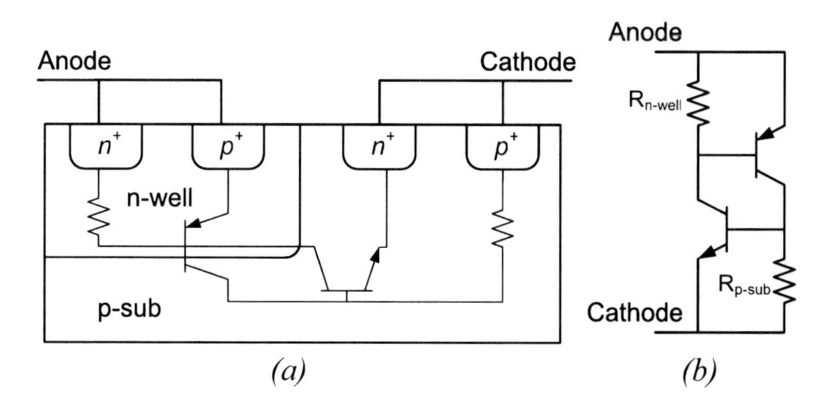

Figure 3-12. Silicon controlled rectifier (a) cross-section (b) equivalent schematic.

bias for the *npn* transistor. Hence, the anode voltage will be reduced to V_h. It can be seen that the *I-V* characteristic of the SCR is similar to that of the GGNMOS.

The value of V_h and V_{t1} of SCR and MOS can be compared based on their operation and structure. In an SCR, breakdown is initiated by the avalanche breakdown of the nwell-psub junction. On the other hand, the breakdown of an NMOS is initiated by n^+-psub avalanche breakdown. In order to compare V_{t1} of NMOS and SCR, avalanche breakdown voltage of nwell-psub junction should be compared to n^+-psub junction. Consider a typical CMOS technology where substrate doping is approximately 5×10^{15} cm^{-3}, well doping is 5×10^{16} cm^{-3} and source/drain doping is 5×10^{20} cm^{-3}. For a GGNMOS, N_D (drain doping) is much more than N_A (substrate doping) while for an SCR, N_A (substrate doping) and N_D (well doping) are in the same order. Referring back to equation (3-2), the breakdown voltage of the NMOS should be much less than the breakdown voltage of the SCR. For the above example, avalanche breakdown voltage for NMOS and SCR is 8 and 20 V respectively. Typical value of V_{t1} in GGNMOS is between 5 and 10 V while V_{t1} for SCR is between 20 and 25 V. The value of V_h for GGNMOS is usually between 3 and 5 V while V_h for SCR is between 1 and 2 V. Low holding voltage of SCR is due to its internal positive feedback. Referring back to Figure 3-12(b), it can be seen that the equivalent schematic of SCR is similar to latch-up structure in CMOS technology. In holding region, both *npn* and *pnp* transistors are in saturation region. Considering that the anode-cathode voltage is equal to V_h, the holding voltage of the SCR shown in Figure 3-12(b) can be easily calculated from the following equation:

$$V_h = V_{EB1} + V_{CE2(sat)}$$

Knowing that V_{EB1} is between 0.7 and 1 V and $V_{CE2(sat)}$ is between 0.3 to 0.6 V, the holding voltage of the SCR is expected to be between 1 and 2 V.

3.3 Low Voltage Triggered SCR (LVTSCR)

As mentioned earlier, the most important drawback of SCR is its high first breakdown voltage which is due to the high avalanche breakdown voltage of nwell-psub junction. In order to lower V_{t1} of the SCR, an n^+ region can be inserted in the boundary of nwell-psub junction so that the breakdown will be initiated by an n^+-p^- junction. This structure is called Medium-Voltage Triggered SCR (MVTSCR) and is shown in Figure 3-13 [13].

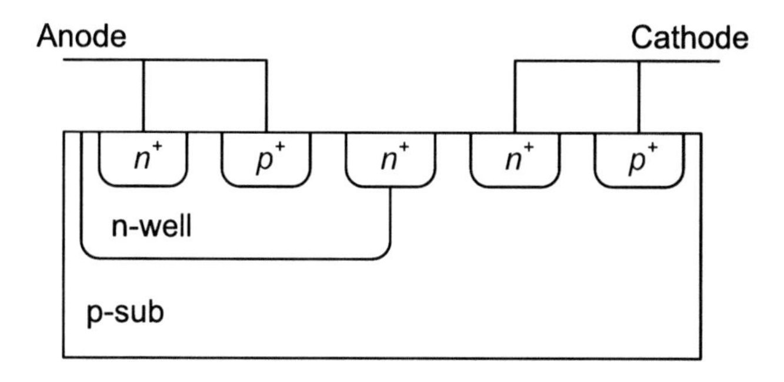

Figure 3-13. Cross section of the MVTSCR.

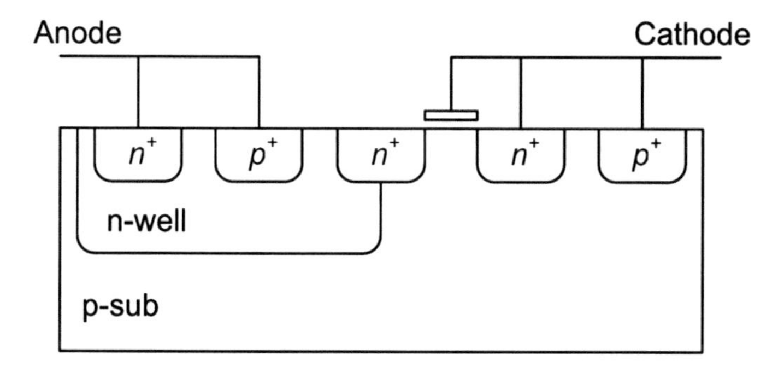

Figure 3-14. Cross section of the LVTSCR.

Breakdown voltage of MVTSCR can be further reduced by inserting a gate electrode between the n^+ cathode and the n^+ added in MVTSCR. This configuration is called Low-Voltage Triggered SCR (LVTSCR) and is shown in Figure 3-14 [14]. The breakdown voltage of LVTSCR is equal to breakdown voltage of its internal NMOS structure. It has been reported the breakdown voltage of LVTSCR is 5 times less than SCR [14].

3.4 Dual SCR

In the previous section, the operation of SCR and NMOS was discussed under PS-mode conditions only. In order to do the protection for NS-mode, complementary of these devices should be added in parallel with the original device. For NMOS, a PMOS transistor is required and for SCR an *npnp* structure. This will double the parasitic capacitance added to the pad. Therefore, based on SCR operation, a new device called dual SCR has been

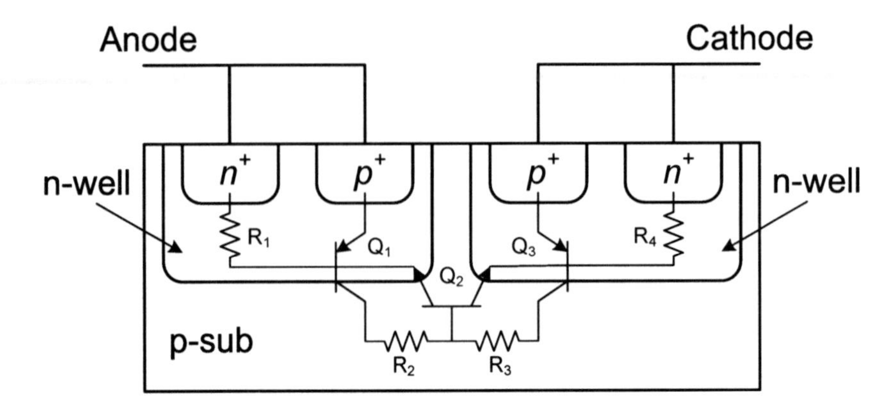

Figure 3-15. Cross section of the dual SCR.

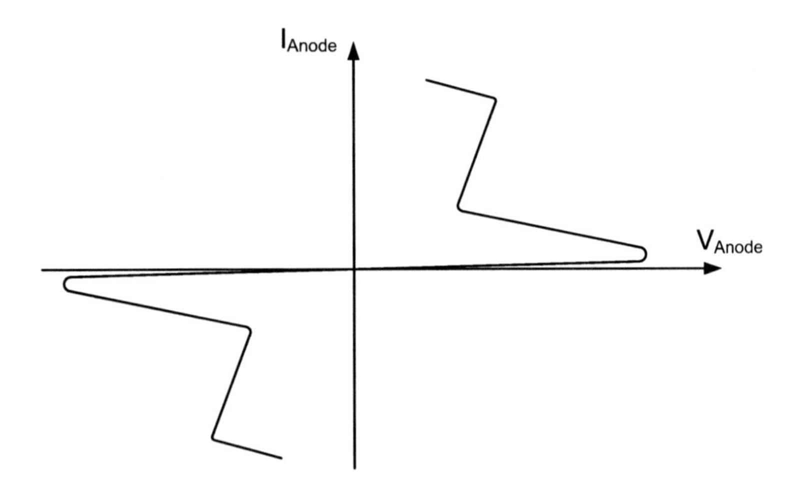

Figure 3-16. I-V characteristic of the dual SCR.

presented which can do the protection for both NS- and PS-modes without adding an extra capacitance [15, 16]. Figure 3-15 shows the cross section of a dual SCR which is a five layer structure.

There are three parasitic bipolar transistors Q_1, Q_2 and Q_3 and four parasitic resistors R_1, R_2, R_3 and R_4 in this structure that are shown in Figure 3-15 as well. For PS- and NS-modes pad is connected to the anode and V_{SS} is connected to the cathode. The path for PS-mode is through the SCR formed by Q_1, Q_2, R_1 and R_3 while the path for NS-mode is through the SCR formed by Q_2, Q_3, R_2 and R_4. Having similar discharge paths for both PS- and NS-modes provides a symmetrical DC characteristic for the dual SCR. Figure 3-16 shows the *I-V* characteristic of the dual SCR.

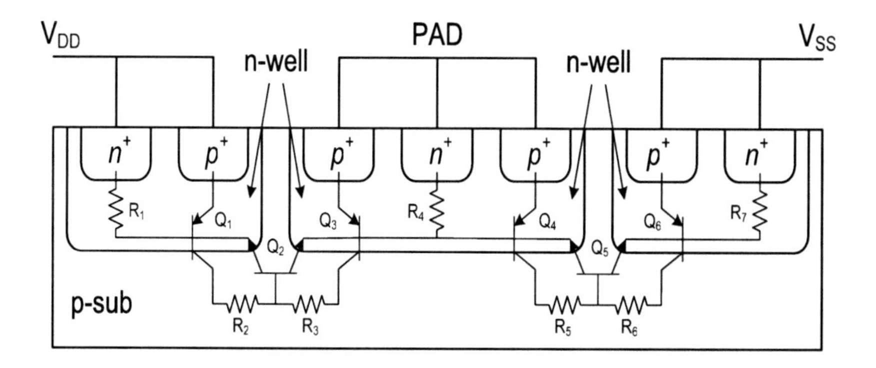

Figure 3-17. Cross section of the all direction SCR.

As mentioned earlier, the main advantage of this technique is smaller parasitic capacitance. This can be verified by comparing the cross section of SCR and dual SCR. It can be seen that in both structures pad is connected to n^+ and p^+ diffusions and therefore their parasitic capacitance are the same. Hence, as one dual SCR or two SCRs are needed to do PS- and NS-mode protection, the parasitic capacitance of dual SCR protection is half of SCR protection.

In order to do the protection for all four zapping modes two dual SCRs are required; one for PS- and NS-modes and one for PD- and ND-modes. To reduce parasitic capacitance further, similar idea can be applied to dual SCR to combine two dual SCRs into one single device. This configuration is called all direction SCR and is shown in Figure 3-17 [16].

It can be seen that this device consists of six parasitic bipolar transistors and eight parasitic resistances. Discharge path for PS-mode is through Q_4, Q_5, R_4 and R_6; for NS-mode is through Q_5, Q_6, R_5 and R_7; for PD-mode is through Q_1, Q_2, R_1 and R_3; for ND-mode is through Q_2, Q_3, R_2 and R_4. It's been reported that by using all-direction SCR total parasitic capacitance of the pad is reduced by close to 8 times [16].

3.5 Gate and Substrate Triggering

As mentioned before, in an ESD protection circuit, the value of V_{t1} should be less than the oxide breakdown voltage. As technology scales down, the oxide breakdown voltage is also decreased. Figure 3-18 shows the gate oxide breakdown voltage for different CMOS technologies [17].

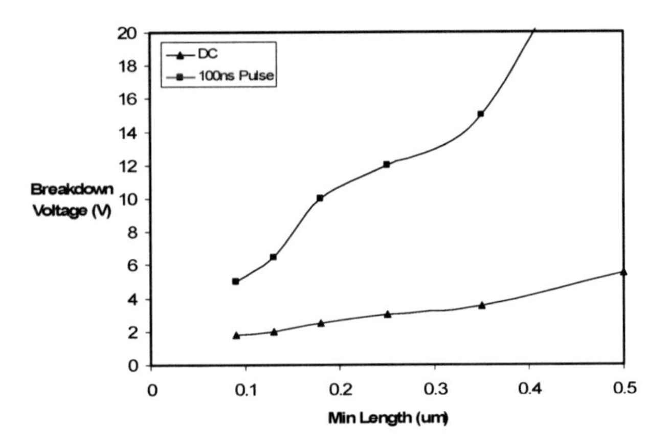

Figure 3-18. Breakdown voltage of thin gate oxide as a function of the CMOS technology generation.

Therefore, in deep submicron CMOS technologies, single ESD devices such as GGNMOS or LVTSCR cannot do the protection by themselves. There are two major circuit techniques to further reduce the triggering voltage of GGNMOS and LVTSCR devices. These techniques are based on the fact that by applying a small voltage to either gate or substrate of GGNMOS or LVTSCR, the triggering voltage is decreased. Gate-triggering technique is based on applying a voltage to the gate while substrate-triggering technique is based on applying a voltage to the substrate. For simplicity, further discussion for gate and substrate-triggering will be based on GGNMOS. Similar discussion can be applied to LVTSCR as well.

In order to understand the impact of biasing the gate or substrate, consider the cross section of the GGNMOS shown in Figure 3-9(a). Applying a small voltage to the gate creates additional current in the substrate due to MOS action. This current will further forward bias the base-emitter junction of the parasitic bipolar transistor. Therefore, the bipolar transistor will turn on with smaller drain voltage, and hence, the triggering voltage will be reduced. Figure 3-19 shows the triggering voltage of an NMOS for different gate voltages. It can be seen that for higher gate voltages, the triggering voltage starts to increase with increasing gate voltage. This is due to the decreased impact ionization rate. In other words, the pinch-off region in the MOS channel disappears and the impact ionization becomes limited by carrier scattering in the inversion region.

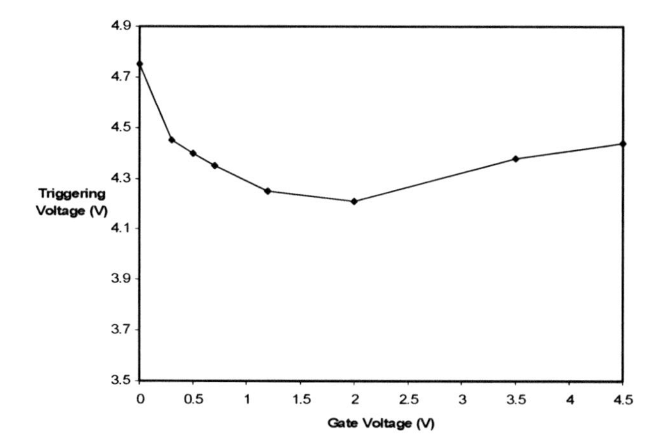

Figure 3-19. Impact of gate voltage on V_{t1}.

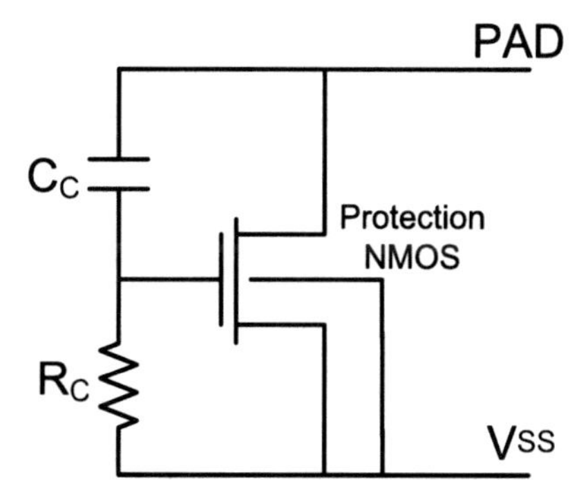

Figure 3-20. Gate-coupled NMOS.

As the ESD event might happen when the IC is not powered up, the gate bias should come from the pad voltage through a coupling circuit. Figure 3-20 shows a simple structure for gate coupling which is called Gate-Coupled NMOS (GCNMOS) [18].

It can be seen that an RC network is used to couple a fraction of ESD charge to the gate of the NMOS transistor. Under ESD conditions, as the pad voltage is increased the gate voltage is increased as well which decreases the triggering voltage. R_C and C_C should be chosen in such a way that under normal operating conditions (when maximum pad voltage is V_{DD}), the gate voltage be less than threshold voltage of the NMOS and under ESD conditions the triggering voltage be close to optimum point in V(gate)–V_{t1} graph (Figure 3-19). An NMOS transistor can be modeled with its gate-source and gate-drain capacitances. Considering relatively constant capacitances, a simple calculation can be done for the values of R_C and C_C. Figure 3-21 shows this simple model.

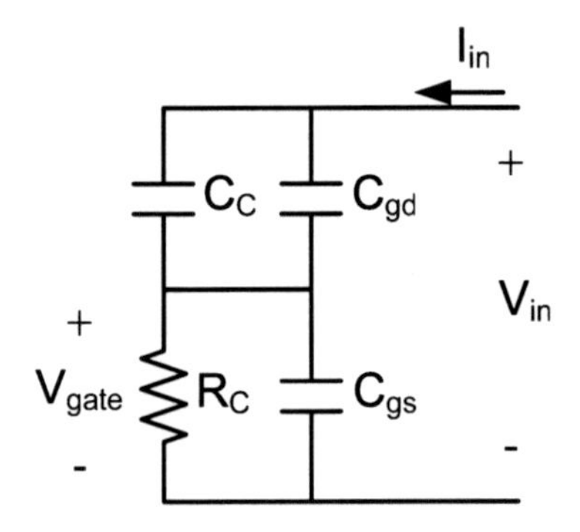

Figure 3-21. A simple model for GCNMOS.

Writing KCL equation for the gate node and ignoring gate-drain capacitance results in the following equations.

$$I_{Cgs} = C_{gs}\frac{dV_{gate}}{dt} = I_{in} - \frac{V_{gate}}{R_C} = C_C\frac{d}{dt}\left(V_{in} - V_{gate}\right) - \frac{V_{gate}}{R_C}$$

$$\frac{dV_{gate}}{dt} + \frac{V_{gate}}{R_C\left(C_C + C_{gs}\right)} = \frac{C_C}{C_C + C_{gs}}\frac{dV_{in}}{dt}$$

The condition for an ESD event is that when the pad voltage changes from 0 to 9 V in 10 ns, the gate voltage should be 1.8 V. For normal operating conditions, when the pad voltage changes from 0 to V_{DD} in 10 ns, the gate voltage should be 0.4 V. Applying these conditions and solving the above differential equation, the following results will be obtained.

$$R_C C_C = 1.9 \times 10^{-9},\ R_C C_{gs} = 1.4 \times 10^{-9}$$

Considering a 100 μm wide NMOS:

$$C_{gs} = C_{ox}\,WL = 150\ \text{fF}$$

$$R_C = 9.4\ \text{k}\Omega,\ C_C = 200\ \text{fF}$$

Usually, final optimizations are carried out with circuit simulations. Using these values for R_C and C_C for circuit simulations, the gate voltage in ESD conditions and normal operating conditions were found to be 1.888 and 0.387 V respectively. These values show good agreement with the design objectives. In the next step this design should be transferred into a device simulator to verify its effectiveness under ESD conditions. Applying a 2 kV HBM stress in Medici the maximum pad voltage is limited to 5.7 V.

It can be seen that using this method, the additional capacitance added to the pad is around 400 fF. This additional capacitance can cause performance degradation especially for high frequency applications.

In order to reduce the additional capacitance required for gate-triggering technique, another gate-triggering method has been proposed which only uses a small coupling NMOS transistor to bias the protection NMOS. Figure 3-22 shows the schematic of this gate triggering technique [19].

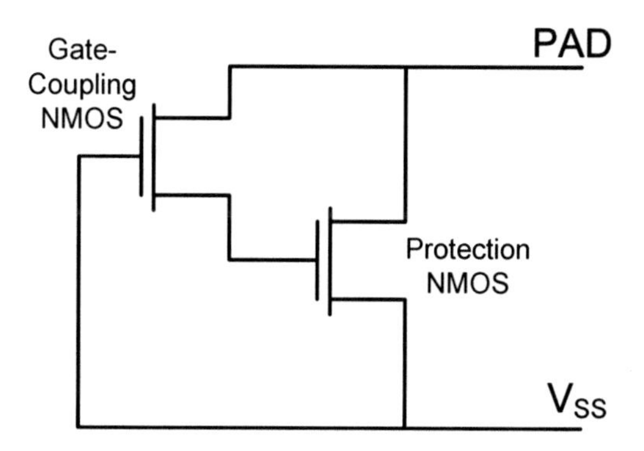

Figure 3-22. Gate triggering technique.

Coupling NMOS should be design to keep the protection NMOS OFF in normal operating conditions and biases the gate under ESD conditions. It's been shown that a 5 μm NMOS can satisfy the design conditions [19]. The only consideration for this design is to make sure the coupling NMOS will not be destroyed under ESD conditions.

It can be seen that gate coupling has positive impact on the first triggering voltage which allows uniform triggering of all fingers and ensures ESD protection in deep submicron technologies. As mentioned in Section 3-1 another important parameter is the second breakdown current (I_{t2}) which determines ESD protection level of the device. Therefore, it is necessary to study the impact of gate-triggering on the second breakdown current. The coupled gate voltage turns on the strong inversion channel of the NMOS. Hence, ESD current will discharge through this region. On the other hand, deep submicron technologies use shallow junction depths and LDD structures which limit the amount of current that can be discharged through this region. Therefore, gate-coupling is less effective in deep submicron technologies. The graph in Figure 3-23 strengthens the above mentioned discussion. The figure shows the impact of the gate voltage on the second breakdown current for a 100 μm transistor [20].

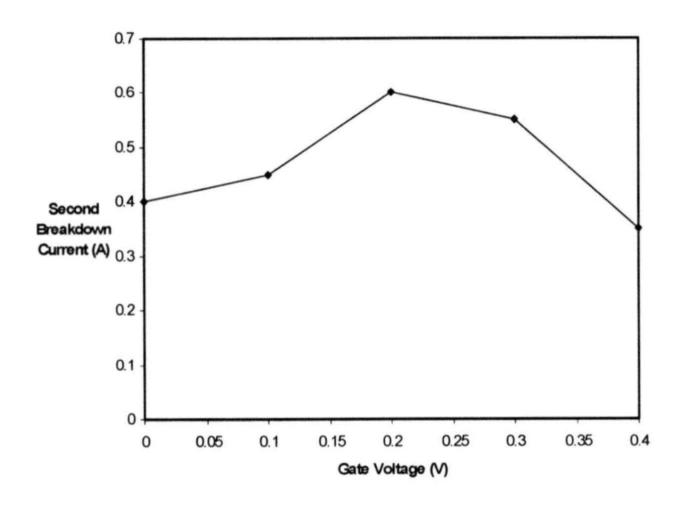

Figure 3-23. Impact of gate bias on I_{t2}.

Similar to biasing the gate, biasing the substrate will decrease the triggering voltage as well. Applying a small voltage to the substrate will increase the base voltage of the parasitic bipolar transistor. This will help forward biasing base-emitter junction of the parasitic bipolar transistor, and hence, the triggering voltage will be reduced. Figure 3-24 shows the triggering voltage of an NMOS transistor for different substrate bias voltages.

There are a number of methods to bias the substrate of an NMOS transistor. Figure 3-25 shows the most common method for substrate triggering which uses another NMOS transistor to bias the substrate [21].

When ESD voltage is applied to the pad the current through substrate triggering NMOS will bias the substrate of the protection NMOS and therefore, reduces its triggering voltage.

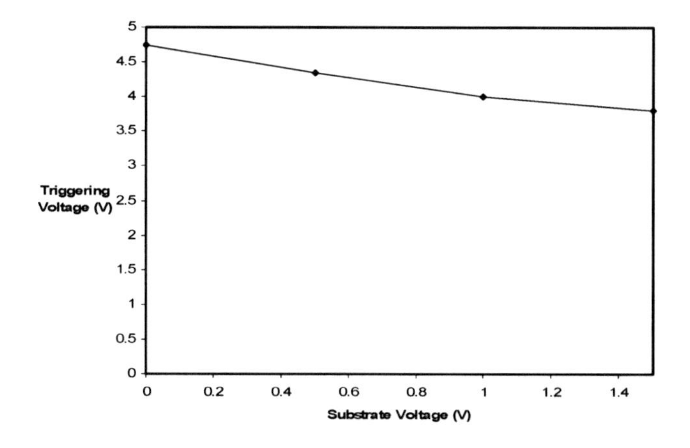

Figure 3-24. Impact of substrate voltage on V_{t1}.

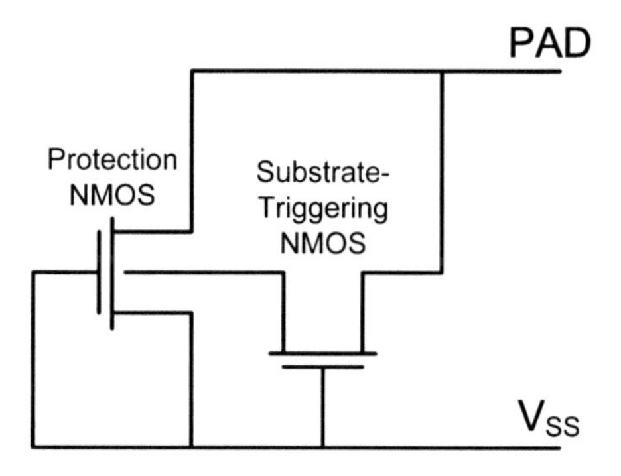

Figure 3-25. Substrate-triggered NMOS (STNMOS).

Unlike gate-triggering, substrate-triggering doesn't have a negative impact on the second breakdown current. The reason is that with substrate triggering, instead of flowing the current through the channel, the current is discharged through a much larger area, substrate. Figure 3-26 shows the impact of substrate-triggering on the second breakdown current [20]. It can be seen that substrate triggering increases I_{t2} and therefore, ESD protection level is increased as well.

In addition to gate-triggering and substrate-triggering, a combination of gate and substrate triggering has been presented which is shown in Figure 3-27 [19].

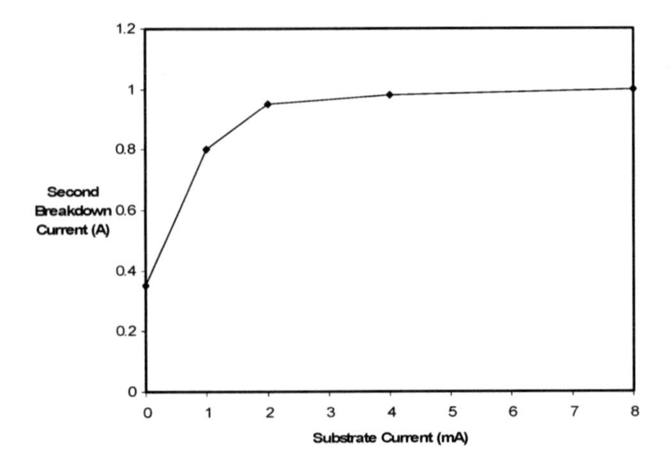

Figure 3-26. Impact of substrate voltage on I_{t2}.

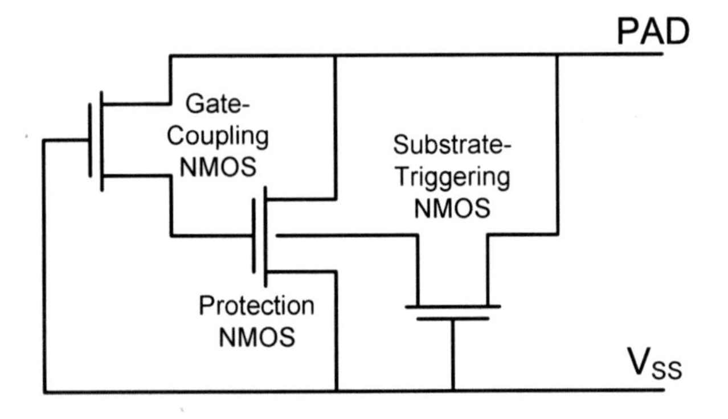

Figure 3-27. Gate-substrate triggering technique.

Compared to either gate or substrate triggering techniques, the combination of gate and substrate triggering can further reduce the triggering voltage. The main disadvantage of this technique is its high leakage current under normal operating conditions. This is due to biasing both gate and substrate of the main transistor. Biasing the gate moves the protection NMOS transistor towards moderate and strong inversion while biasing the substrate reduces the threshold voltage of the transistor. This leakage for a 0.18 μm CMOS technology is in the milli-amp range.

In order to reduce the leakage, another transistor can be used to tie the gate of the protection NMOS to ground under normal operating conditions. Figure 3-28 shows the schematic of the low-leakage gate-substrate triggering technique [22].

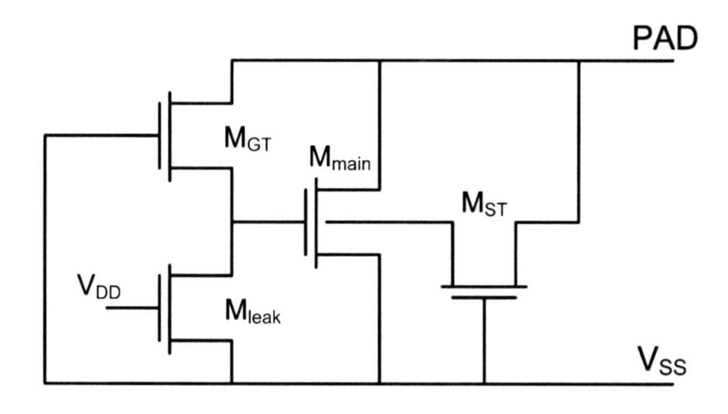

Figure 3-28. Low-leakage gate-substrate triggering NMOS.

Under normal operating conditions transistor M_{leak} is ON which connects the gate of M_{main} to ground. Under ESD conditions, the circuit is not powered and V_{DD} is floating. Hence, the circuit in Figure 3-28 becomes a regular gate-substrate triggered NMOS. This technique reduces leakage by six orders of magnitude is a 0.18 μm CMOS technology.

As mentioned earlier, gate and substrate triggering can be applied to LVTSCR in a similar way.

4. LATCH-UP IN ESD PROTECTION DEVICES

In Section 3 ESD protection devices were discussed and compared based on their first breakdown voltage and second breakdown current. First breakdown voltage is designed to be less than the oxide breakdown voltage while second breakdown current should be as high as possible to increase the ESD protection level. It was shown that SCR based devices have much higher I_{t2} compared to MOS based devices. Therefore SCR family is more widely used in ESD protection applications. On the other hand, there is one major drawback in SCR family of devices. As mentioned in Section 3.2, unlike MOS based devices, holding voltage of SCR family devices is very low and is between 1 to 2 V. As the holding voltage is less than the supply voltage, under normal operating conditions a small current in the substrate can increase the SCR current beyond the holding current and trigger the SCR. This phenomenon is called latch-up. Latch-up is often destructive, hence must be avoided at all costs.

Besides normal operating conditions, semiconductor devices are often tested under stressed conditions such as burn-in as well. These tests are done to perform a reliability test on VLSI circuits by monitoring weak defects such as weak gate oxides. Usually these defects are detected by accelerating their failure under high temperature (≈125°C) and high voltage (V_{DD} + 30%) conditions. In practice, for a chip with 1.5 V power supply, 2.5 V spikes were observed on input pins of Field Programmable Gate Arrays (FPGAs) due to the electrical overstress in burn-in ovens [23]. Therefore, in order to avoid latch-up under burn-in conditions, which results in yield loss [24], holding voltage of ESD protection devices should be at least 50% higher than the supply voltage.

The above discussion reveals the importance of making ESD protection devices immune to latch-up. In addition to increasing the holding voltage, increasing the holding current can prevent latch-up under normal operating conditions as well. Higher holding current reduces the chance for noise currents to sustain latch-up. In the following subsections, circuit techniques are discussed to increase the holding voltage and holding current along with their state of the art implementations.

4.1 Increasing the Holding Voltage

In processes with epitaxial substrate, holding voltage can be increased by increasing the anode to cathode spacing. However, this method doesn't work for bulk CMOS processes. To overcome this problem in bulk CMOS processes, cascoding of SCR devices is suggested by M. D. Ker et al. [25]. Figure 3-29(a) shows three LVTSCRs that are cascaded in order to increase the overall holding voltage of the protection circuit. The overall holding voltage is the sum of the holding voltage of individual LVTSCRs. As a result, the overall holding voltage can be designed by changing the number of LVTSCR devices. Although this method increases the holding voltage, unfortunately it increases the first breakdown voltage as well. Therefore, another technique such as gate-substrate triggering should be applied to LVTSCRs to reduce the overall triggering voltage below the oxide breakdown voltage. In addition, this method increases turn on resistance of the overall circuit and hence, wider LVTSCRs are needed to maintain the original ESD protection level, which may increase the parasitic capacitance associated with the LVTSCR.

This method can be modified by cascading one LVTSCR with a stack of diodes [39]. This implementation is shown in Figure 3-29(b). However, LVSTSCR with stack of diodes may offer marginally better implementation.

It can be seen that there are a number of limitations for cascoding method. In order to explore other options, let's refer back to the equivalent circuit of an SCR shown in Figure 3-30(a). As discussed in Section 3.2, after

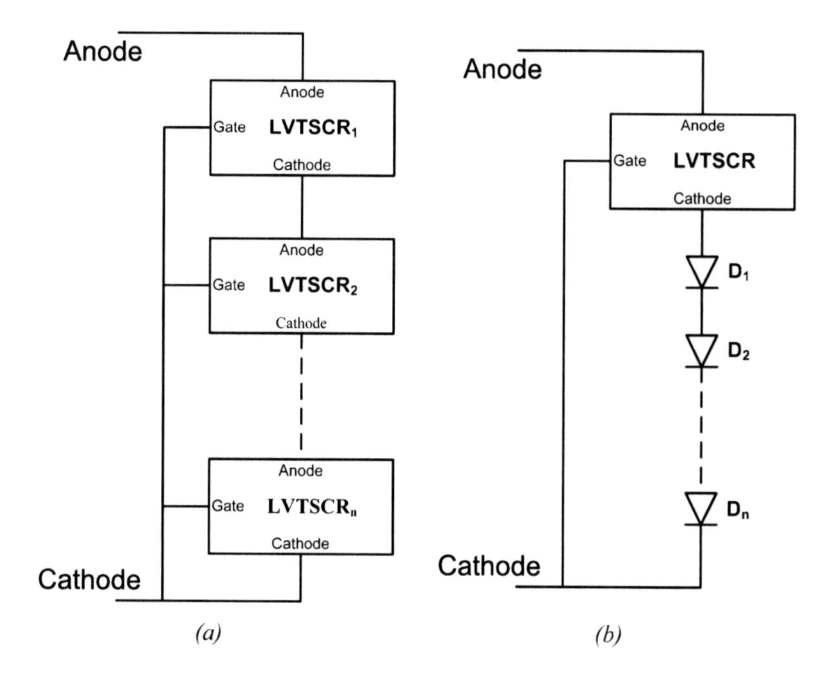

Figure 3-29. Increasing V_h (a) cascoding LVTSCRs (b) cascading LVTSCR with diodes.

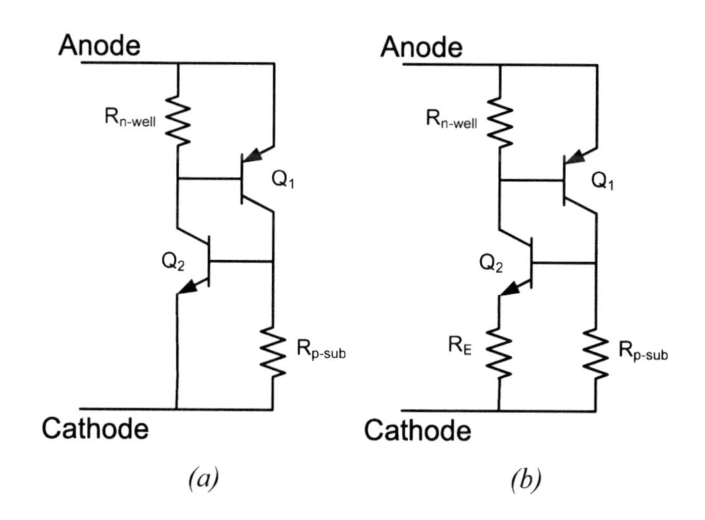

Figure 3-30. Equivalent circuit (a) SCR (b) high V_h SCR.

SCR triggers and both bipolar transistors turn on, the anode voltage is reduced to a value that keeps the bipolar transistors on. This voltage is known as the holding voltage (V_h). In order to increase the holding voltage, the voltage needed to keep one of the bipolar transistors on should be increased. The simplest method is to decrease either $R_{n\text{-well}}$ or $R_{p\text{-sub}}$. As a result, more ESD voltage is needed to generate enough voltage across $R_{n\text{-well}}$ or $R_{p\text{-sub}}$ to keep the bipolar transistors on and therefore, holding voltage is increased. This requires to increase the doping of well or substrate. As doping profile of well and substrate are process dependent parameters, circuit designer cannot change their value and therefore, this method is not a practical solution to increase V_h.

Instead of decreasing well or substrate resistance, one can increase the required voltage across $R_{n\text{-well}}$ or $R_{p\text{-sub}}$ by inserting a resistor in series with emitter of one of the bipolar transistors (Figure 3-30(b)) [28]. The holding voltage of the new SCR can be easily calculated. The holding voltage of the SCR in Figure 3-30(a) equals to.

$$V_h = V_{EB1} + V_{CE2(sat)}$$

Writing a similar KVL equation and neglecting the base current of the bipolar transistors the holding voltage of the high V_h SCR equals to.

$$V_h = V_{EB1} + V_{CE2(sat)} + I_{E2}R_E = V_{EB1} + V_{CE2(sat)} + \frac{V_{EB1}R_E}{R_{n-well}}$$

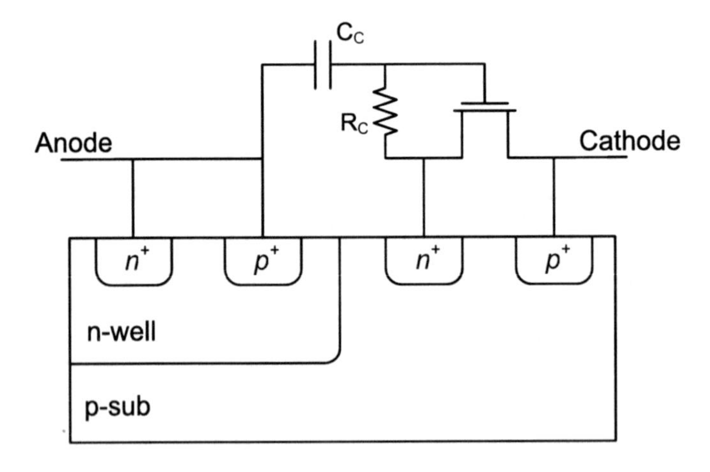

Figure 3-31. Cross section of the high V_h SCR.

It can be seen that by changing the value of R_E, holding voltage of the SCR can be optimized. This resistor can be implemented using poly resistor, diode or MOS transistor. Figure 3-31 shows the cross section of the high V_hSCR where an NMOS is used to implement the resistor [28]. It has been reported that the high V_hSCR shown in Figure 3-31 increases the holding voltage by 2X [28].

4.2 Increasing the Holding Current

As mentioned earlier, latch-up immunity can be achieved by increasing the holding current as well [25, 26]. Considering the equivalent circuit of the SCR shown in Figure 3-30(a), it has been shown that the holding current can be calculated from the following equation [27]:

$$I_h = \frac{\beta_1(\beta_2+1)}{\beta_1\beta_2-1}\frac{V_{EB1}}{R_{n-well}} + \frac{\beta_2(\beta_1+1)}{\beta_1\beta_2-1}\frac{V_{BE2}}{R_{p-sub}}$$

Similar to Section 4.1 a simpler equation can be found by neglecting the base current of the bipolar transistors:

$$I_h = \frac{V_{EB1}}{R_{n-well}} + \frac{V_{BE2}}{R_{p-sub}}$$

In order to increase the holding current another current path is added in the SCR structure. Figure 3-32 shows both the equivalent circuit and the

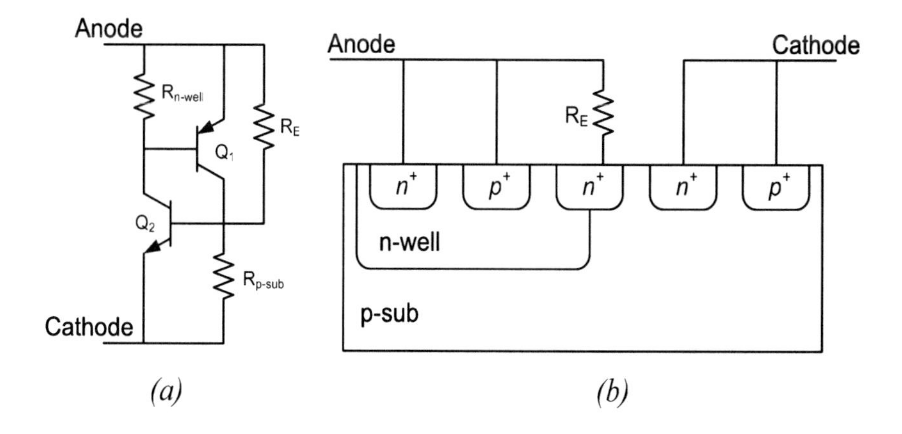

Figure 3-32. High I_h SCR (a) equivalent circuit (b) cross section.

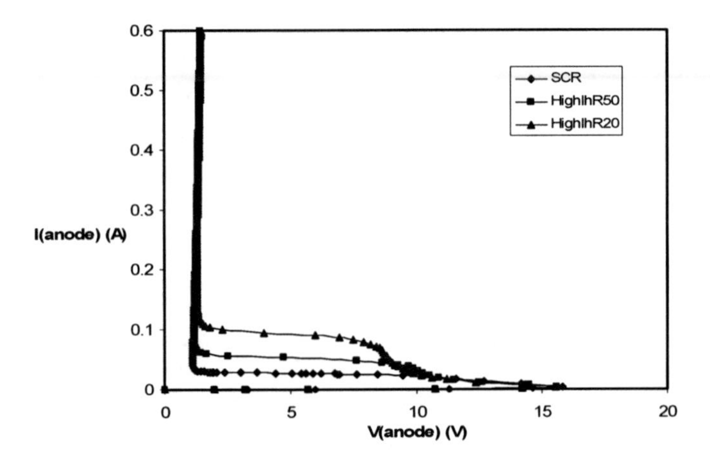

Figure 3-33. Impact of R on holding current.

cross section of an SCR with an additional current path using a simple resistor. The holding current of the high I_hSCR can be calculated from the following equations:

$$I_h = \frac{V_{EB1}}{R_{n-well}} + \frac{V_{BE2}}{R_{p-sub}} + I_{R_E} = \frac{V_{EB1}}{R_{n-well}} + \frac{V_{BE2}}{R_{p-sub}} + \frac{V_{EC1(sat)}}{R_E}$$

Therefore, the increase in the holding current is proportional to the conductance of R_E. Figure 3-33 shows the *I-V* characteristic of high I_hSCR for different values of R_E.

It can be seen that by reducing the value of R_E, the holding current can increase significantly. This resistor can be implemented using either poly resistor or forward biased diode. Similar results are obtained by applying this method to LVTSCR.

5. ESD DEVICES UNDER STRESS CONDITIONS: BURN-IN

As mentioned earlier, burn-in is an important test that is performed on VLSI products. It involves high temperature and high voltage test. Therefore, high holding voltage is a requirement for ESD protection devices. Discussions on Section 4 were based on increasing holding voltage/current of ESD protection devices. Similar discussion can be applied to elevated voltage under burn-in test. Therefore, ESD protection device should be designed to target a holding voltage higher than maximum voltage under burn-in conditions.

On the other hand, burn-in test involves elevated temperatures as well. Therefore, the impact of temperature on holding voltage/current of semiconductor devices should be studied. Burn-in ovens often use temperatures up to 125°C or 398 K. Therefore, in this section the impact of the temperatures up to 400 K on the holding voltage/current of ESD protection devices will be discussed. Again as SCR devices are more vulnerable to latch-up the discussions in this section are for SCR family ESD protection devices.

As discussed earlier, in an SCR and in holding region both parasitic bipolar transistors are in saturation region and holding voltage is the sum of V_{EB1} and $V_{CE2(sat)}$ (refer to Figure 3-30(a)). By increasing temperature V_{EB1} which equals to $V_{EB(on)}$ decreases while $V_{CE2(sat)}$ remains almost constant. Therefore, holding voltage of SCR reduces in high temperatures, which makes them more vulnerable to latch-up in burn-in conditions. In order to study the impact of high temperature on holding voltage of SCR family devices, device level simulations should be done on major SCR family structures: LVTSCR, LVTSCR with gate triggering, LVTSCR with substrate triggering, LVTSCR with gate and substrate triggering and finally high V_h LVTSCR. Figure 3-34 shows the value of V_h for different structures in 300–400 K temperature range [28].

It can be seen that a linear dependency on temperature exists for holding voltage in 300–400 K range. The notations in Figure 3-34 show the reduction of holding voltage in percentage when the ambient temperature increases from

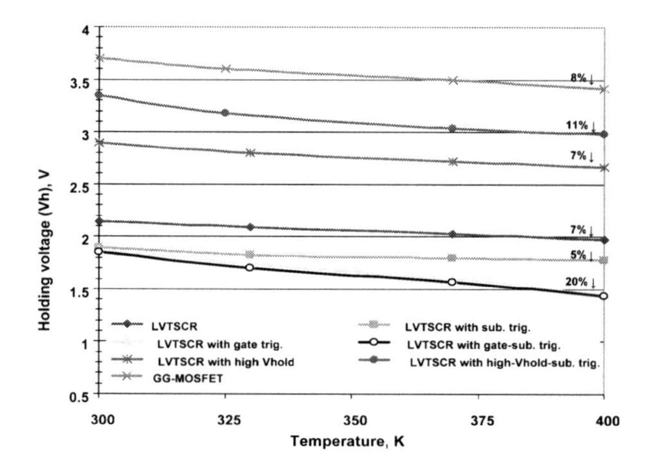

Figure 3-34. Impact of temperature on V_h.

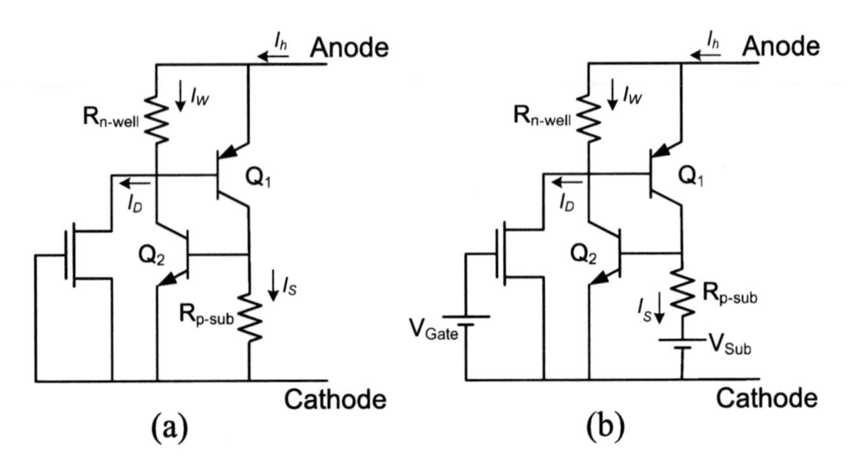

Figure 3-35. Analysis of holding current (a) LVTSCR (b) LVTSCR with gate/substrate triggering.

room temperature to burn-in temperature. Comparing different structures it can be seen that LVTSCR with gate triggering has the holding voltage degradation by 4X stronger than the LVTSCR with substrate triggering. Hence, for burn-in conditions and to reduce the triggering voltage, the substrate triggering technique is the best option. In addition, LVTSCR with high V_h exhibit low temperature dependency as well.

Similar conclusions can be driven for holding current [28]. Consider the model for LVTSCR shown in Figure 3-35(a). Writing the current equations:

$$I_h = I_{E1} + I_W = (\beta_p + 1)I_{B1} + I_W \tag{3-11}$$

$$I_W + I_{B1} = I_D + I_{C2} = I_D\big|_{V_{GS}=0} + \beta_n I_{B2} \tag{3-12}$$

$$I_S + I_{B2} = I_{C1} = \beta_p I_{B1} \tag{3-13}$$

Finding the value of I_{B1} from (3-9) and (3-10) and substituting it into (3-8) the holding current can be calculated.

$$I_h = \frac{\beta_p(\beta_n + 1)I_W + \beta_n(\beta_p + 1)I_S}{\beta_n\beta_p - 1} - \frac{\beta_p + 1}{\beta_n\beta_p - 1} I_D\big|_{V_{GS}=0} \tag{3-14}$$

The first term in (3-11) is the holding current of SCR and for simplicity (3-11) can be written in the following form.

$$I_{h(LVTSCR)} = I_{h(SCR)} - \frac{\beta_p + 1}{\beta_n\beta_p - 1} I_D\big|_{V_{GS}=0} \tag{3-15}$$

In order to use this model for gate triggering and substrate triggering conditions, two bias voltages are added to the model, which represent gate and substrate bias of the LVTSCR. This model is shown in Figure 3-35(b). According to the new model, the equation (3-12) can be modified to cover the impact of gate triggering and substrate triggering on the holding current of LVTSCR.

$$I_{h(GT-LVTSCR)} = I_{h(SCR)} - \frac{\beta_p + 1}{\beta_n \beta_p - 1} I_D \Big|_{V_{gate}}$$

$$I_{h(ST-LVTSCR)} = I_{h(SCR)} - \frac{\beta_p + 1}{\beta_n \beta_p - 1} \frac{V_{sub}}{R_{p-sub}} - \frac{\beta_p + 1}{\beta_n \beta_p - 1} I_D \Big|_{V_{GS}=0}$$

In order to model the temperature dependency of holding current, gain of bipolar transistors (β_n and β_p), well and substrate resistances ($R_{n\text{-well}}$ and $R_{p\text{-sub}}$), and MOSFET current (I_D) should be determined in terms of temperature. In other words, to predict variations of holding current under burn-in conditions, the temperature dependency of mobility (μ_n and μ_p), current gain of bipolar transistor (β_n and β_p) and MOSFET threshold voltage (V_T) should be derived. It has been reported that V_T has a linear dependency on temperature. For 0.18 μm CMOS technology, dV_T/dT is 0.6 mV/°C [29]. Current gain of bipolar transistor has positive temperature dependence. It's been shown that holding current of all the above LVTSCR-based devices has negative temperature dependence. Similar to holding voltage trend, gate triggering shows higher reduction in holding current compared to substrate triggering which makes substrate triggering a preferred choice under burn-in conditions. This similarity is expected as the LVTSCR shows a resistive behavior in holding region.

6. FAILURE CRITERIA OF ESD DEVICES

In order to determine the ESD protection level of semiconductor devices, failure analysis should be done on semiconductor devices. There are three main failure criteria for ESD devices: the second breakdown current (I_{t2}) [30], the leakage current (I_{off}) measured after ESD stress [31] and the failure temperature or the current at failure temperature [32].

6.1 I_{t2} Current Criterion

As discussed in Section 3, SCR and MOS have a snapback behavior under ESD conditions. Second breakdown region is when thermal damage is occurring in the device. Therefore, I_{t2}, which can be measured easily, can be used as an indicator of the maximum allowed current. However, the actual damaging ESD current usually drives devices well beyond the trigger point of the second breakdown [32]. Hence, this criterion gives the pessimistic estimation of ESD robustness.

6.2 Leakage Current (I_{off}) Criterion

The I_{t2} corresponds to the current that flows through the device just prior to the device failure. However, in many cases, the device exhibits an increased drain-to-substrate leakage current even before it reaches the second breakdown point in the *I-V* curve. Because of this, the leakage current is used as a monitor of ESD robustness. Typically, the I_{off} criterion is based on the specified value of leakage current after the ESD stress. If the leakage current is higher than the specified value, which is typically 10 nA, 100 nA or 1 μA, then the device is considered failed according to this failure criterion.

For example, the failure criterion during the product qualification and/or screening tests is typically stated as 1 μA [31]. However, the experimental data show that before the ESD device destruction by current filament a soft leakage current occurs in NMOS because of gate oxide degradation. The typical range of soft leakage current in ESD NMOS is 400–750 nA [33]. Hence, the I_{off} criterion can not be always used as a reliable indicator of actual ESD device destruction.

6.3 Failure Temperature Criterion

Destruction of an ESD device occurs at the threshold voltage, at which the maximum temperature reaches the melting point of silicon (1,412°C) [32] (typically in the gate-to-drain overlap region) or the melting point of metallization (660°C for aluminum based metallization and 1,034°C for copper based metallization) [34]. With respect to this ESD robustness criterion, the current I_{1412} is the ESD threshold current (and $\approx 1{,}500 \times I_{1,412}$ is the ESD threshold voltage) that will raise the temperature at the hottest spots to 1,412°C [32], the melting temperature of silicon. Depending on the failure mechanism, which is typical for a given ESD device, the I_{1034} or I_{660} current limit can be also used. The principle disadvantage of this criterion is the necessity to measure the temperature rise in the device during an ESD event on a nanosecond time scale. Several techniques were developed for the measurement of temperature dynamics under nanosecond high current stress, such as IR thermal interferometry [35], optical pyrometry [36] and backside laser interferometry [37].

7. SUMMARY

In this chapter the main semiconductor devices that can provide ESD protection circuit in CMOS technology are discussed. These devices are divided into two main categories: snapback; and non-snapback devices. The *p-n* junction diode, zener diode and polysilicon diode are the most popular non-snapback devices, while snapback devices are mainly based on MOSFETs and SCRs. SCRs have the advantage of having relatively smaller size, lower on-state impedance and lower capacitance for a given ESD protection level. On the other hand, historically their disadvantages have been higher triggering voltages and higher turn-on time [40, 41], which makes them sensitive to CDM damage. Overall, diode and SCR-based devices have the highest ESD protection level per unit area and are suitable for high speed applications.

ESD engineers should pay attention to latch up if they intend to use SCR based ESD protection circuits. In Section 4 latch-up in SCR-based devices was discussed along with two main solutions to design a latch-up immune protection circuit. Finally, the impact of burn-in test on ESD robustness of deep submicron ICs is also discussed.

REFERENCES

[1] S. M. Sze, *Semiconductor devices physics and technology*, Wiley, New York, 2002.

[2] J. W. Miller, "Application and process dependent ESD design strategy," Tutorial in *EOS/ESD Symposium*, 2003.

[3] E. Rosenbaum, S. Hyvonen, "On-chip ESD protection for RF I/Os: devices, circuits and models," *IEEE Int. Symp. Cir. and Sys.*, pp. 1202–1205, 2005.

[4] Y. Blecher and R. Fried, "Zener substrate triggering for CMOS ESD protection devices," *Electronic Letters*, vol. 32, No. 22, pp. 2102–2103, 1996.

[5] Y. Tsividis, *Operation and modeling of the MOS transistor*, Oxford University Press, New York, 2006.

[6] L. Luh, J. Choma, and J. Draper, "A zener-diode-activated ESD protection circuit for submicron CMOS processes," *IEEE Int. Symp. Cir. and Sys.*, pp. 65–68, 2000.

[7] C. Y. Chang and M. D. Ker, "On-chip ESD protection design for GHz RF integrated circuits by using polysilicon diodes in sub-quarter-micron CMOS process," *IEEE Int. Symp. VLSI Tech. Sys. App.*, pp. 240–243, 2001.

[8] G. Chen, et al., "RF characterization of ESD protection structures," *IEEE RF Int. Cir. Symp.*, pp. 379–382, 2004.

[9] F. C. Hsu, P. K. Ko, S. Tam, C. Hu, and R. S. Muller, "An analytical breakdown model for short-channel MOSFET's," *IEEE Trans. on Electron Dev.*, vol. ED-29, No. 11, 1982.

[10] D. Krakauer and K. Mistry, "ESD protection in a 3.3 V sub-micron silicided CMOS technology," *Proc. EOS/ESD Symposium*, pp. 250–257, 1992.

[11] G. Notermans, "On the use of n-well resistors for uniform triggering of ESD protection elements," *Proc. EOS/ESD Symposium*, pp. 221–229, 1997.

[12] K. G. Verhaege and C. Russ, "Wafer cost reduction through design of high performance fully silicided ESD devices," *Proc. EOS/ESD Symposium*, pp. 18–28, 2000.
[13] C. Duvvury and R. Rountree, "A synthesis of ESD input protection circuit," *Proc. EOS/ESD Symposium*, pp. 69–84, 1991.
[14] A. Chatterjee and T. Polgreen, "A low-voltage triggering SCR for on-chip ESD protection at output and input pads," *IEEE Elec. Dev. Letters*, pp. 21–22, 1991.
[15] A. Z. H. Wang and C. H. Tsay, "On a dual-polarity on-chip electrostatic discharge protection structure," *IEEE Trans. Elec. Dev.*, vol. 48, No. 5, pp. 978–984, 2001.
[16] H. Feng, K. Gong, and A. Z. Wang, "A comparison study of ESD protection for RFIC's: performance vs parasitics," *IEEE RF Int. Cir. Symp.*, pp. 235–238, 2000.
[17] H. Gossner, "ESD protection for the deep sub-micron regime – a challenge for design methodology," *Proc Int. Conf VLSI Des.*, pp. 809–818, 2004.
[18] R. Merrill and E. Issaq, "ESD design methodology," *Proc. EOS/ESD Symposium*, pp. 233–237, 1993.
[19] O. Semenov, H. Sarbishaei, V. Axelrad, and M. Sachdev, "Novel gate and substrate triggering techniques for deep submicron ESD protection devices," *Microelectronics Journal*, vol. 37, No. 6, pp. 526–533, 2006.
[20] T. Y. Chen and M. D. Ker, "Investigation of the gate-driven effect and substrate-triggered effect on ESD robustness of CMOS devices," *IEEE Trans. on Dev. Materials Reliability*, vol. 1, No. 4, pp. 190–203, 2002.
[21] M. D. Ker, T. Y. Chen, and C. Y. Wu, "ESD protection design in a 0.18 μm silicide CMOS technology by using a substrate-triggered technique," *IEEE Int. Symp. Cir. Sys.*, pp. 754–757, 2001.
[22] H. Sarbishaei, O. Semenov, and M. Sachdev, "Optimizing circuit performance and ESD protection for high-speed differential I/Os," *submitted to Custom Int. Cir. Conf.*, 2007.
[23] M. Sawant, et al., "Post programming burn in (PPBI) for RT54SX-S and A54SX-A ACTEL FPGAs," http://www.actel.com/products/aero/ppbi_rev5minal.pdf
[24] C. Y. Chiang, "Latch-up at RAM control circuitry", *Proc. of IPFA*, pp. 250–253, 1997.
[25] M. D. Ker and H. H. Chang, "How to safely apply the LVTSCR for CMOS whole-chip ESD protection without being accidentally triggered on," *Proc. EOS/ESD Symposium*, pp. 72–85, 1998.
[26] M. P. J. Mergens, C. C. Russ, K. G. Verhaege, J. Armer, P. C. Jozwaik, and R. Mohn, "High holding current SCRs (HHI-SCR) for ESD protection and latch-up immune IC operation," *EOS/ESD Symposium*, paper 1A3, 2002.
[27] F. S. Shoucair, "High-temperature latchup characteristics in VLSI CMOS circuits," *IEEE Trans. on Electron Dev.*, vol. 35, No. 12, pp. 2424–2426, 1988.
[28] O. Semenov, H. Sarbishaei, and M. Sachdev, "Analysis and design of LVTSCR-based EOS/ESD protection circuits for burn-in environment," *Proc. of the Int. Symp. Quality Electron Design*, pp. 427–432, 2005.
[29] O. Semenov, et al., "Leakage current in sub-quarter micron MOSFET: a perspective on stresses delta I_{DDQ} testing," *Journal of Electronic Testing: Theory and Application*, vol. 19, No. 3, pp. 341–352, 2003.
[30] A. Amerasekera and C. Duvvury, "The impact of technology scaling on ESD robustness and protection circuit design," *IEEE Trans.on Component, Packaging and Manufacturing Tech.*, Part A, vol. 18, No. 2, pp. 314–320, 1995.
[31] C. Meneghesso, S. Santirosi, E. Novarini, C. Contiero, and E. Zanoni, "ESD robustness of smart-power protection structures evaluated by means of HBM and TLP tests," *IEEE IRPS*, pp. 270–275, 2000.
[32] D. L. Lin, "ESD sensitivity and VLSI technology trends: thermal breakdown and dielectric breakdown," *Proc. EOS/ESD Symposium*, pp. 73–81, 1993.

[33] T. Wadano, "Study of the soft leakage induced ESD on LDD transistor," *Microelectronics Reliability*, vol. 36, No. 11/12, pp. 1707–1710, 1996.
[34] S. H. Voldman, "The impact of technology scaling on ESD robustness of aluminum and copper interconnects in advanced semiconductor technologies," *IEEE Trans. on Component, Packaging and Manufacturing Tech.*, Part C, vol. 21, No. 4, pp. 265–277, 1998.
[35] K. Murakami, K. Takita, and K. Masuda, "Measurement of lattice temperature during pulsed laser-annealing by time-dependent optical reflectivity," *Japanese Journal of Applied Physics*, vol. 20, No. 12, pp. L867–L870, 1981.
[36] I. P. Herman, "Real time optical thermometry during semiconductor processing," *IEEE J. on Selected Topics in Quantum Electron.*, vol. 1, No. 4, pp. 1047–1053, 1995.
[37] C. Furbock, N. Seliger, D. Pogany, M. Litzenberger, E. Gornik, M. Stecher, H. Gossner, and W. Werener, "Backside laser probe characterization of thermal effects during high current stress in smart power ESD protection devices," *Proc. of IEDM*, pp. 691–694, 1998.
[38] P. R. Gray, P. J. Hurst, S. H. Lewis, and R. G. Meyer, *Analysis and Design of Analog Integrated Circuits*, Wiley, New York, 2001.
[39] M. D. Ker and K. C. Hsu, "Latchup-free ESD protection design with complementary substrate-triggered SCR devices," *IEEE J. Solid State Cir.*, vol. 38, No. 8, pp. 1380–1392, 2003.
[40] P. A. Juliano and E. Rosenbaum, "A novel SCR macromodel for ESD circuit simulation," *Int. Elec. Dev. Meeting*, pp. 14.3.1–14.3.4, 2001.
[41] M. D. Ker and H. K Chun, "Native-NMOS-triggered SCR (NANSCR) for ESD protection in 0.13 μm CMOS integrated circuits," *Int. Rel. Phys. Symp.*, pp. 381–386, 2004.

Chapter 4

CIRCUIT DESIGN CONCEPTS FOR ESD PROTECTION

1. INTRODUCTION

The internal ESD protection requires the placement of adequate on-chip protection devices on the I/O and on the power supply pins to reliably bypass the ESD energy before it can damage the sensitive circuits. The on-chip protection scheme should have an explicit and robust path for the ESD currents to flow between any pair of pins. In general, pad protection networks shunt I/O pins to the ground or V_{DD} bus under stress events. For each input pin, a dedicated protection network, that is completely passive under normal operating conditions, has to be added. For each output pin, the ESD protection level is determined by the intrinsic robustness of the output buffer transistors plus that of the dedicated protection devices. A good protection element should minimize the nominal performance and/or voltage degradation to the I/O circuit due to its insertion and provide a low-impedance shunt path for the ESD current. The protection element must be capable of handling multiple ESD events without itself being destroyed. It should also not interfere with the I/O circuit during its normal operation. Hence, a perfect protection device should have the following characteristic:

- Very low on-resistance
- Triggering voltage should be above the worst case operating supply voltage (V_{DD} + 10%)
- Almost instantaneous turn-on time

O. Semenov et al., ESD Protection Device and Circuit Design for Advanced CMOS Technologies, 85–116.

- Very high energy handling capability
- Only trigger during ESD events, not during normal operation
- Very low parasitics to minimize performance degradation of I/O circuit
- Consumes small area

Very low on-resistance allows it to shunt large amount of current with no voltage rise from an ohmic voltage drop. It is clear that a real ESD protect-tion device can not have all of these characteristics, but these criteria provide a list of optimizations and compromises to be struck when the protection circuit is designed.

The ESD devices used for I/O protection in deep submicron CMOS technologies can be combined in two groups: non-snapback protection devices and snapback protection devices. The first group includes diodes, zener diodes and punch through transient voltage suppressors (TVS). The second group typically includes MOSFETs, lateral double diffused MOS (LDMOS) transistors, silicon-controlled rectifiers (SCR) and low voltage triggered SCRs (LVTSCR).

This chapter describes the general principles used to develop the ESD protection circuits that not only meet the ESD objectives but also meet the functional objectives of the I/O. At the beginning of this chapter the different ESD protection circuits for different applications are considered. The second part of the chapter describes the design flow of ESD protection circuits. It covers the ESD device simulations and calibration, mixed-mode (device-circuit) simulations and chip level ESD simulations. The design of special test structures for ESD network verification and ESD measurement concepts are also considered in this chapter.

2. ESD PROTECTION NETWORKS

Today's ICs pack the complex operations of multiple functional sub-systems, such us analog, RF circuits, memory, digital signal processing (DSP) units, input/output controls, onto one single "system on a chip" (SoC) design. Since each functional sub-system may have a unique supply voltage requirement, multiple power supplies are distributed across the IC and form complex internal power networks. Both the heterogeneous nature of SoCs and multiple supply domains are resulting in complex on-chip ESD protection networks and they are typically optimized for the given application of the IC.

Different ESD protection elements should be combined to form a protection network. One well-known I/O ESD protection circuit is a pi-network as shown in Figure 4-1 [1]. During the ESD stress, the secondary ESD clamp triggers first and then the voltage drop through the resistor is increased and it triggers the primary ESD clamp. Subsequently, most of ESD energy is bypassed through the primary ESD clamp. The major advantages of the pi-protection network are the voltage division and safer triggering of ESD devices. The first advantage ensures the lower voltage stress in the internal node, and the second one gives lower trigger voltage in comparison with using the primary ESD clamp only.

For the digital ICs the multi ESD bus strategy is popular as shown in Figure 4-2 [1]. In this case, each I/O pad has multi ESD clamps with respect to ESD buses, which are typically ground bus and power bus. The power line should also have the power ESD clamp. For the ESD current path between

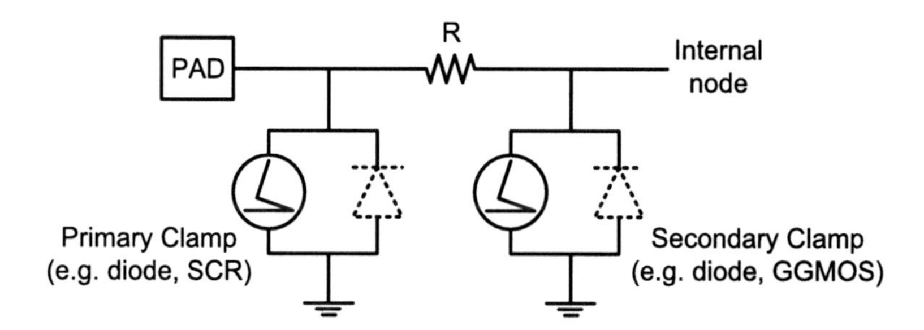

Figure 4-1. The pi-ESD protection network for I/O. (Adapted from [1].)

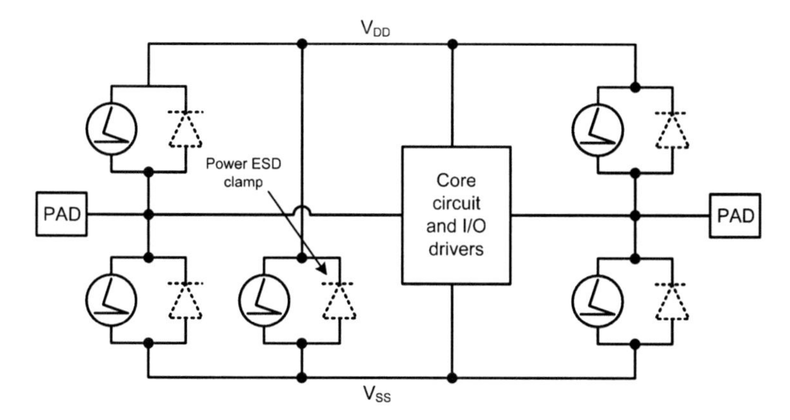

Figure 4-2. The general ESD protection network of whole chip.

pad and the V_{DD} or the V_{SS}, simple diodes in a forward biased configuration can be used if the V_{DD} is equal or higher than the pad voltage. If the pad voltage is higher than V_{DD} then a suitable diode chain can be used. The ESD current path between V_{DD} and V_{SS} is provided by the core clamp. This device is normally off during the circuit operation. During the ESD event it has low impedance path. The current path between V_{SS} and pad is provided by the forward biased diode. Here one diode is generally sufficient. Typically, diodes provide the current path from the pad to V_{DD} and NMOS core clamp provides the current path between V_{DD} and V_{SS} [2].

Some circuits must work in an environment where input voltage levels exceed the supply lines. Examples of these are multiplexers and switches. For instance, the inputs can be specified with a 24 V over-voltage rating, even though the supplies are rated at 12 V. Having inputs exceeding the supply lines becomes challenging, because typical protection techniques, as shown in Figure 4-2, can not be used. One technique, which can solve this issue, is to allow the inputs to be tied to an isolated bus on chip [3]. This technique is illustrated in Figure 4-3. When the ESD/EOS event is happened on I/O pad, its energy is dissipated through the forward biased I/O ESD diode, floating rail #1, supply clamp, floating rail #2 and ESD diode connected to the grounded (V-) pad or other grounded I/O pad, respectively.

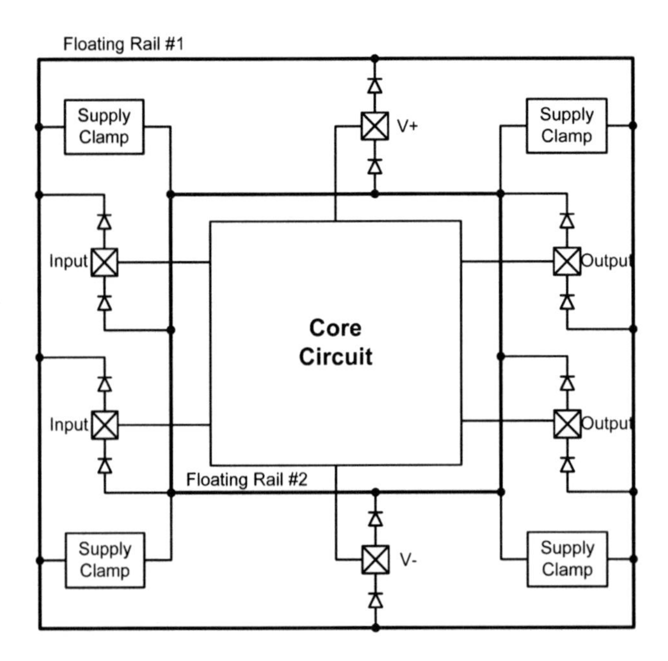

Figure 4-3. ESD protection network for input voltages that exceed supply levels. (Adapted from [3].)

The supply busses and ESD busses are typically used to provide a low ohmic path between any couple of pins during the ESD event. In the simplest form the supply busses are organized as a ring at the perimeter of the IC for the IO supply and a power mesh in the centre of the IC for the core supply. Any resistance along this path will count for the ESD design window of the core protection. This resistance can be controlled by the width of the metal busses and the distance between the supply pads and connections on the package substrate. Typical values for wirebond ICs with an acceptable density of supply pads are 1–2 Ω [4]. For the 2 kV HBM ESD stress, which is equivalent to a discharge current of 1.3 A, this leads to a voltage drop of 1.3–2.6 V.

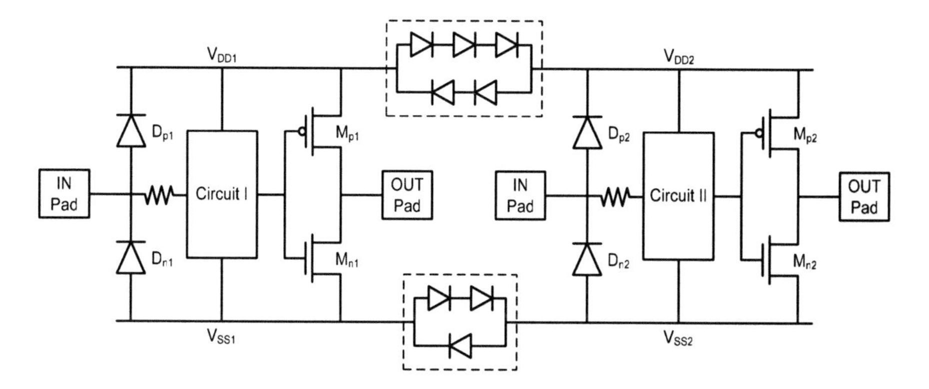

Figure 4-4. ESD protection network for coupling between two power domains in mixed-mode ICs. (Adapted from [6].)

In most modern designs a simple power domain arrangement is not any longer used. Especially for low power designs, for example mobile applications, there are a larger number of power domains both for core and for IO circuitry, which should be independently powered. Similarly, a high-integration giga-scale SoC chips often have multiple and separated power lines, since the analog, RF, memory and digital circuits have different voltage and ESD noise requirements [5, 6]. In addition, the V_{SS} lines are often decoupled to avoid cross-talk between the domains. The typical ESD protection network for multi-V_{DD} ICs is shown in Figure 4-4, where the bi-directional diode strings are connected between the V_{DD1} and V_{DD2}, and between the V_{SS1} and V_{SS2}. The number of diodes in the diode string between the separated power lines depends on the voltage level or the noise level between these power lines. The diode strings are designed to conduct the ESD current between the separated power lines to avoid the ESD damage of internal circuits, when the IC is under the ESD stress. At normal operating conditions, the diode string is designed to block the voltage or noise between the separated power lines.

Diode-based ESD power networks are commonly used in many applications due to easy implementation and area efficiency. However, diode-based ESD networks are not favored in view of noise margins needed for mixed mode (digital-analog) circuit operations. In a single n-well CMOS process, the stringing of diodes causes formation of parasitic PNP bipolar transistors and during PNP transistor operation some fraction of the emitter current can sink into the substrate. In the noisy environment due to digital switching, the delay elements within the voltage-controlled oscillator (VCO) become sensitive to the Power-Ground (PG) noise. A change in the supply-ground coupling of a VCO shifts the oscillation frequency. The magnitude of the error depends on the delay element's sensitivity to PG noise and the loop bandwidth. Ground noise coupling through the protection diodes results in fluctuations in the control voltage and on the local ground of the VCO.

In addition, if the CMOS IC has several separated power pines to supply different circuit blocks, the ESD current will be discharged through one or more diode strings connected between the separated power lines [6]. The multiple diode-strings in the ESD current discharging paths lead to a higher path resistance and longer delay to bypass the ESD current away from the internal circuits of the CMOS IC with separated power pins. It may result into the damage of internal circuits. Hence, the diode-based ESD power protection network can be not suitable for the CMOS ICs, which have

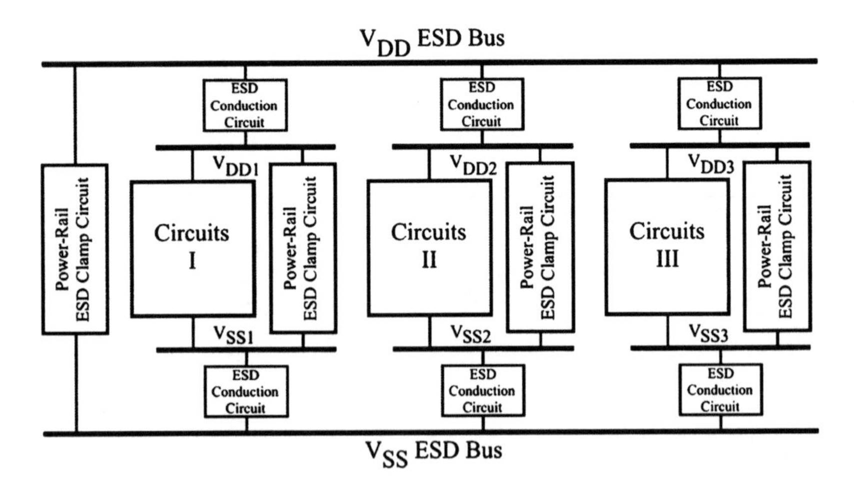

Figure 4-5. Whole-chip ESD protection design with two ESD buses for mixed-voltage pins. (Adapted from [7].)

much more than 2–3 separated power pins. To overcome the limitations of conventional diode-based ESD power protection network, the ESD buses ESD protection scheme was developed [7, 8]. By using ESD busses, the ESD stress current can be quickly discharged far away from the internal circuits or interface circuits of CMOS ICs. A simplified version of whole-chip ESD protection design with two ESD busses is shown in Figure 4-5. In this figure, the ESD conduction circuits can be the bi-directional ESD cells, such as dual-mode SCR or two NMOS triggered SCR devices [8]. Typical power rail ESD clamp circuits include gate grounded NMOS transistors (GG-MOSFETs), RC triggered PMOSFETs and NMOSFETs, stacked diode strings, and NMOS triggered SCR devices.

3. DISTRIBUTED ESD PROTECTION NETWORKS

The previously mentioned ESD protection networks have some limitations and disadvantages. The first primary disadvantage is that power ESD clamps require a large layout area. Two or more rail clamps, the size of an on-chip wire bond pad, are often needed to dissipate the required ESD current. Since these power clamp circuits are quite large, they must be placed wherever space is available in the pad ring. This is typically in power and ground pad cells or in large spacer cells between pads. This results in the second primary disadvantage of active MOSFET-based ESD protection networks. In many chip applications, large banks of tightly packed I/O pads must be placed, offering no room for large ESD rail clamp circuits. In these "pad limited" designs, I/O ESD robustness is reduced with increasing distance from the remote rail clamps, simply due to bus resistance, which is typically 1–2 Ω. For the 2 kV HBM ESD stress, which is equivalent to a discharge current of 1.3 A, this leads to a voltage drop of 1.3–2.6 V. As a result, each new chip design should be optimized in respect to the size and location of the power clamps and the power bus resistances in order to protect I/O pads.

To overcome disadvantages of ESD protection networks based on the large transient triggered MOSFET rail clamp circuits, which described in details in Chapter 5, distributed active MOSFET power clamp ESD networks were developed [9, 10]. In an ideal on-chip ESD network, rail clamps should be placed in close proximity to all external pads to minimize the impact of power rail resistance. In a distributed ESD network, rail clamp NMOS transistors are distributed in each I/O pad cell, while the primary clamp triggering elements remain remotely placed in power and ground pad cells. In other words, small rail clamp circuits and buffer circuits are placed in each I/O pad cell, and the primary RC trigger is placed in remote locations, like the power and ground pad cells. This distributed rail clamp

ESD network is illustrated in Figure 4-6. A small NMOS transistor, a buffer circuit, and a small capacitor are placed in each I/O pad cell. In each power or ground pad cell a complete large rail clamp circuit, including large NMOS transistor, buffer, and RC circuit, is placed. A key feature of this approach is that the output of RC trigger in each of power pad cells is connected to the small clamps in each I/O pad cell by a narrow metal bus (ESD_RC). Note that both the small capacitors (C2), placed in each I/O pad cell, and the ESD_RC bus resistance supplement the remote large capacitor C1 and resistor R1 in the distributed rail clamp operation. The shown in Figure 4-6 ESD network there is no problem with delayed turn-on of the distributed clamps since, for positive V_{DD} to V_{SS} zaps, each clamp is triggered on locally, by capacitors C1 or C2. The time-out function, on the other hand, is controlled remotely, via the resistor R1. For this reason, this design can tolerate a narrow, somewhat resistive ESD_RC bus. A key feature of the distributed approach is that for any stressed pad in a group of I/Os, the ESD stress does not fall only on the small clamp local to the stressed pad. This local clamp and its neighbors on each side work in parallel to safely dissipate the ESD current. Due to this distributed protection, the clamps in each I/O may be sized quite small to minimize their layout area.

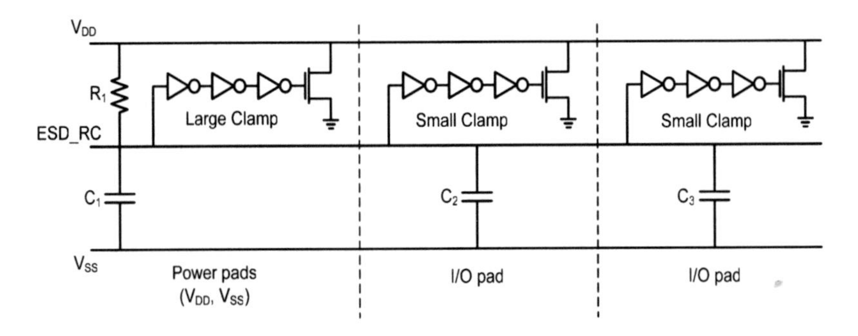

Figure 4-6. Distributed rail clamp ESD network with remote trigger. (Adapted from [10].)

The important issue of distributed rail clamp ESD networks is the optimization of ESD element sizing in each I/O pad and large clamp circuits. In distributed ESD network design, the large rail clamps are only needed to terminate a group of distributed small rail clamps. Therefore, the I/O pad ESD devices should be sized so that in the center of a very large I/O group, the distributed small clamps provide full protection, with no assistance from the distant large clamps. It was found, that as the number of I/O pads in particular group increases, the optimum ESD device sizes are reduced and eventually saturated. For example, for the relatively high bus resistance

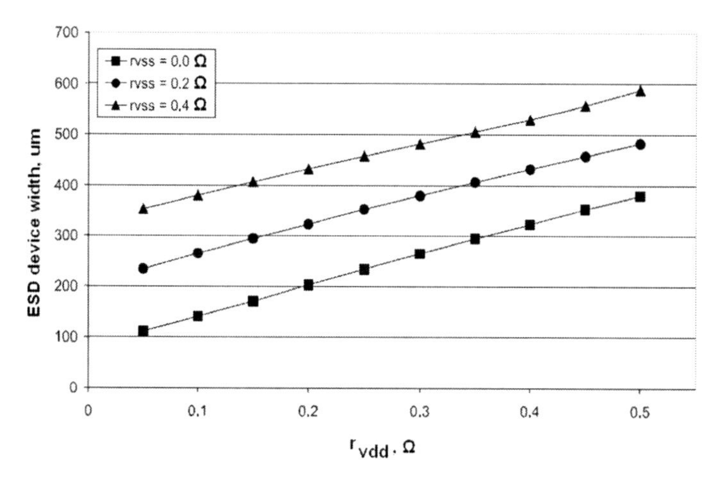

Figure 4-7. Optimized widths of small rail clamp NMOSFET for 0.25 µm CMOS technology. (Adapted from [10].)

networks, the ESD NMOSFET sizes are effectively saturated at about 21 pads. In case of low bus resistance networks, the ESD NMOSFET sizes are effectively saturated at about 60 pads [10]. The calculated optimal I/O pad ESD device sizes for the center I/O in the unterminated 61 I/O pads group, for the full range of bus resistance values, are shown in Figure 4-7.

In this figure, r_{vdd} and r_{vss} values represent the incremental (pad-to-pad) V_{DD} and V_{SS} bus resistances, respectively. A second set of optimizations should be performed to determine the minimum width of the large rail clamp NMOSFET in the power pads at both ends of the I/O group to effectively terminate the network of distributed small rail clamps. One possible approach to do this analysis is to optimize the large rail clamp NMOSFET width based on the ESD stress applied to a single I/O pad terminated on both sides with large rail clamp circuits. The small rail clamp NMOSFET width in the I/O pad can be fixed at the values determined from the previous optimizations. Using this procedure, it was calculated the optimum large rail clamp NMOSFET width for the full range of bus resistance values [10]. In Figure 4-8, the optimum ESD device width is plotted as a function of r_{vdd} for the three different r_{vss} values. From this figure, we can conclude that the optimized large rail clamp width is slightly increased with decreasing of bus resistance. The optimum large rail clamp NMOSFET width falls within a narrow 1,800–2,000 µm range. The distributed rail clamp ESD protection network was first implemented in 0.25 µm CMOS technology and currently is widely used in Freescale Semiconductor for different products.

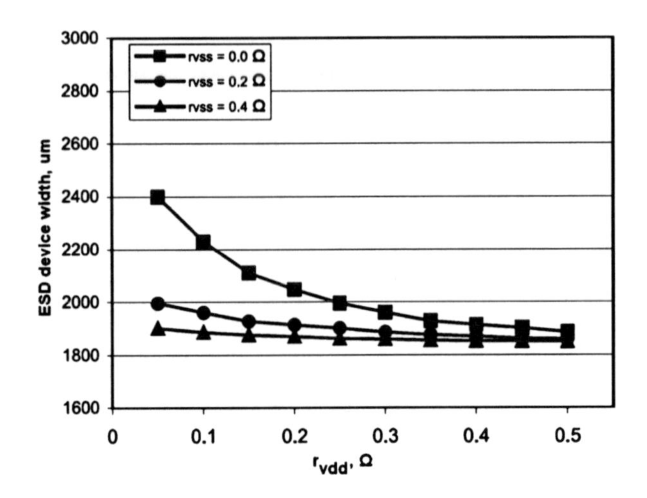

Figure 4-8. Optimized widths of large rail clamp NMOSFET for 0.25 μm CMOS technology. (Adapted from [10].)

3.1 Distributed Boosted ESD Networks

It is well known that for a given drain-to-source voltage (V_{ds}), the conductance of an NMOSFET increases as the gate-to-source voltage (V_{gs}) is increased. In the boosted ESD rail configuration [11], it was found a way to apply a larger V_{gs} voltage to the gate of ESD MOSFET (M0) in a large clamp circuit during the ESD event. Figure 4-9 shows a general schematic of the boosted active MOSFET rail clamp configuration [12]. Note that there are two changes in boosted configuration in comparison with the non-boosted configuration. First, a very narrow Boost bus, with two parasitic bus resistors $R_1 = 20\ \Omega$, was added. The Boost bus serves as a power supply bus for the rail clamp trigger circuit. Note, that the drain of M0 remains connected to V_{DD}. Second, a relatively small (20 μm widths) VPNP protection device (A2) in diode configuration was added in each I/O pad cell. The A2 device couples ESD current onto the Boost bus from the stressed I/O pad. The primary ESD current path is through A1, M0 and B protection devices and there is a very little current flow through A2 device in the stressed I/O pad. Therefore, the voltage drop on the Boost bus at stressed I/O is only about a diode drop below the I/O pad voltage. In addition, an insignificant voltage drop is seen across the two Boost bus resistors R1. It was shown by Stockinger et al. [12] that in boosted ESD rail protection network, the V_{gs} voltage of rail clamp NMOSFET (M0) is ~2 × higher than the V_{gs} voltage of rail clamp protection device in non-boosted ESD railprotection network at the same V_{ds} voltage. As a result, the rail clamp NMOSFET (M0) may

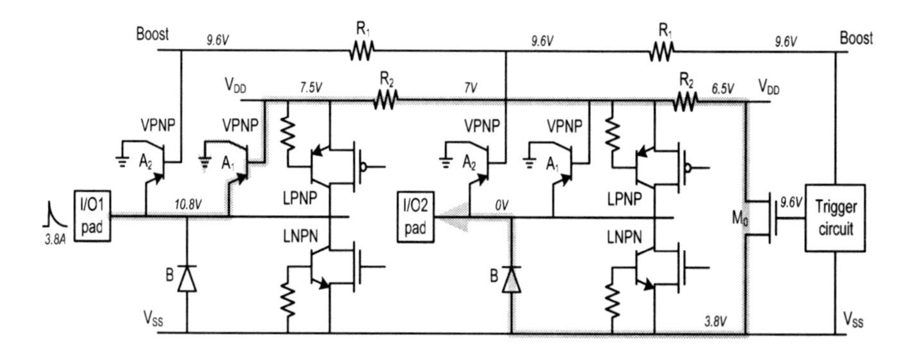

Figure 4-9. Boosted rail clamp ESD protection network. (Adapted from [12].)

pass significantly higher ESD current. For the same ESD robustness, M0 transistor can be smaller by 2.3 × in boosted ESD rail protection network in comparison with non-boosted ESD rail protection network.

An obvious concern with the boosted ESD rail clamp configuration is that the clamp NMOS M0 may be at increased risk of damage under the $V_{gs} > V_{ds}$ boosted bias conditions. At an ESD event, the boosted ESD rail protection network should be optimized to avoid the snapback operating mode of M0 transistor and V_{gs} voltage should be not higher than the gate oxide breakdown voltage for the given CMOS technology.

In case of distributed boosted ESD networks shown in Figure 4-10, the single large boosted active MOSFET rail clamp (M0) can be split into multiple, much smaller clamp NMOS devices M1, connected in parallel, and distributed in each of the I/O and power supply pad cells for efficient, uniform ESD protection. The incremental parasitic bus resistances R1, R2, R3, and R4 for each bus are shown between each pad cell. In the boosted and distributed ESD rail clamp network shown in Figure 4-10, the ESD bus serves as the high current rail clamp anode bus and the V_{SS} bus serves as the cathode bus. Therefore, it is important that these two buses should be as wide as possible to ensure the low resistance to ESD currents flowing around the chip periphery. The Boost and Trigger buses do not move significant current during an ESD event, and may, therefore, be much more narrow and resistive. Each I/O pad cell in Figure 4-10 contains a small rail clamp NMOSFET M1 and ESD diodes A1, A2, and B. A1 and A2 represent the emitter-base junction diodes of the VPNP devices A1 and A2 shown in Figure 4-9. The V_{SS} pad cell contains the same ESD elements as the I/O pad cells and also provides a convenient location for a remote ESD rail clamp

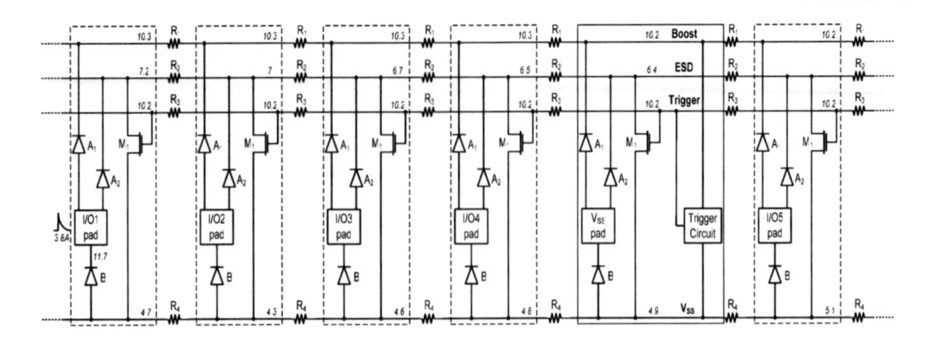

Figure 4-10. Optimized boosted and distributed ESD network. (Adapted from [12].) Nodal voltages are shown for 3.8 A ESD event applied at I/O1 with I/O2 grounded.

trigger circuit. This trigger circuit is defined as remote because, in addition to the local NMOSFET M1, it also drives, via the Trigger bus, NMOSFET devices M1 beyond the V_{SS} pad cell. With remote trigger circuits, no local trigger circuits are needed in the I/O pad cells, saving significant layout area. Since there is very little IR voltage drop along the Boost and Trigger buses, trigger circuits may be placed in some distance from the stressed pad and some distance from the distributed clamp NMOS devices, which they drive. In a non-boosted network, large IR voltage drops along the ESD bus make it almost impossible to effectively use remote trigger circuits in this manner.

The ESD bus may be either a floating bus, not directly connected to any external power supply, or may also serve as a positive power supply (i.e. V_{DD}) bus. For ICs, where the ESD bus is also a V_{DD} bus connected to the external V_{DD} pad, V_{DD} zaps positive to grounded I/O can be a problem. The V_{DD} pad has the same design as the V_{SS} pad shown in Figure 4-10. Under this zap condition, the large voltage drop across A2 can be seen at the trigger circuit that can be damaged, since there is no voltage drop at A1. In addition, it may not adequately protect the PMOSET of output buffer in the grounded pad [12]. It is possible that the PMOSFET in the grounded I/O pad may be stressed such that its drain voltage will be higher than the first breakdown voltage (V_{t1}). To solve this problem, the large clamp NMOS devices (M0) can be placed in each V_{DD} pad cell, to locally compensate for the ESD clamp performance reduction with hot zaps at V_{DD}. In this case, the large clamp NMOS devices (M0) can terminate a group of distributed small rail clamp NMOS devices (M1), and can locally protect I/O buffers from the ESD event at V_{DD} pad.

4. CIRCUIT DESIGN FLOW FOR ESD

The goal of the ESD protection strategy is to avoid damage both due to the high discharge currents and due to the extreme over-voltage in ICs. In sub-100 nm CMOS technologies even an over-voltage of 2–3 V can cause damage to the core devices. Hence, it is become a challenging task to simultaneously optimize a given ESD strategy under conflicting and restrictive constraints of area, performance and ESD requirements. The restrictive nature of these circuit constraints are reflected by the narrowing of ESD design window which will be discussed subsequently in this section. In this context, device simulations provide a rare insight into ESD device behavior under high current and voltage conditions. Based on the process description, the detailed analysis of the device behavior can be performed and design improvements can be carried out in relatively short time.

In most modern designs a simple power domain arrangement is not any longer used. For example, there are a large number of power domains for core as well as for I/O circuitry that should be independently powered in low power, mobile applications. Needless to say, there are several independent V_{DD} supplies. In addition, the V_{SS} lines are decoupled to avoid the cross-talk between the domains. As a result realizing effective and efficient ESD becomes a challenging task. *"How the ESD discharge current can safely pass through an arbitrary combination of pins and power domains?"* becomes an important consideration. The evaluation only pass/fail HBM/MM conditions for the ESD test structures, as it was in past, can not be applied for VLSI designs in sub-100 nm technologies. Due to the significantly thinner gate oxide of core transistors in these technologies, the CDM stress conditions become critical. The CDM ESD failures occur typically in cores instead of IO blocks as it was observed in previous CMOS technologies for HBM/MM ESD events. Often, CDM failures are placed in interface regions of multi-power domains. Since in CDM stress charges are localized inside of the cores, it's extremely difficult to predict their discharge path. Hence, the ESD design strategy must consider the real discharge conditions at the chip level. However, this requires the development of a new IC level ESD simulation approaches.

4.1 Device Simulation and Calibration

One has freedom to choose from a large selection of ESD protection device types. However, it depends on the process technology and the circuit application. Nevertheless, all protection devices must satisfy certain conditions concerning their *I-V* characteristics, which are described by the ESD

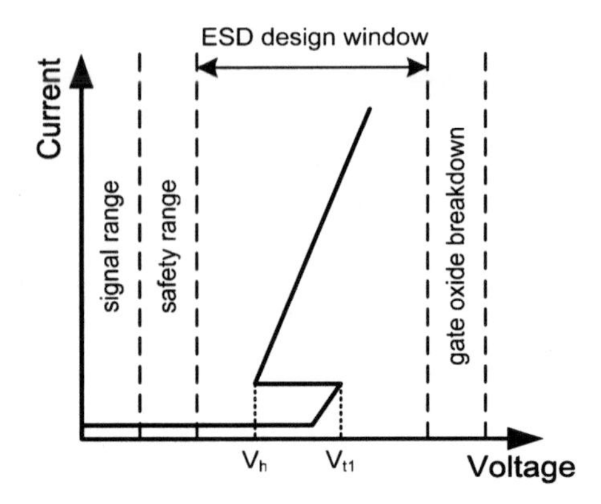

Figure 4-11. ESD design window between the VDD and the gate oxide breakdown voltage of the core circuit.

design window shown in Figure 4-11. Many of the ESD protection devices operate in a so-called snap-back mode. The protection device should not limit the normal operation of ICs. It means that its breakdown voltage (V_{t1}) must be above the signal range (V_{DD}) plus some noise margin and the holding voltage (V_h) should be also higher than the V_{DD} to avoid latch-up susceptibility under normal operating conditions.

With CMOS technology scaling, the gate oxide thickness of MOSFETs in core circuit is drastically reduced. As a result, the gate oxide breakdown voltage is also diminished as depicted in Figure 4-12 and the width of the ESD window for the core protection is shrunk to ~3–4 V for the 100 nm

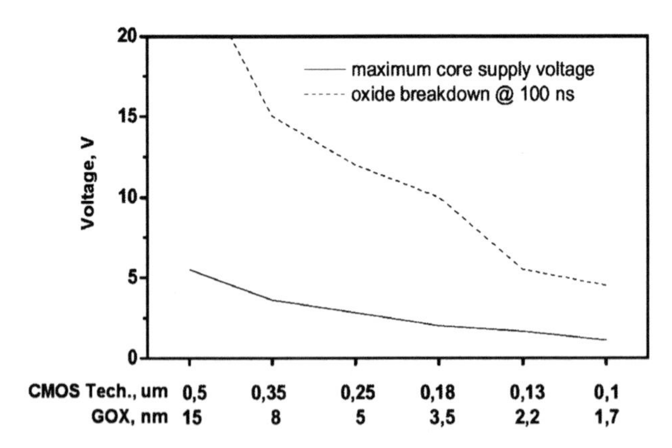

Figure 4-12. Breakdown voltage of gate oxide for 100 ns duration stress pulses and supply voltage as a function of CMOS technology generation. (Adapted from [4].)

CMOS technology node [4]. This ESD window must guarantee the safe operation of core circuit at the full operational temperature range. Needless to say, it requires a balancing of the protection device features, behavior under nominal conditions, and a careful analysis concerning impact of process fluctuations.

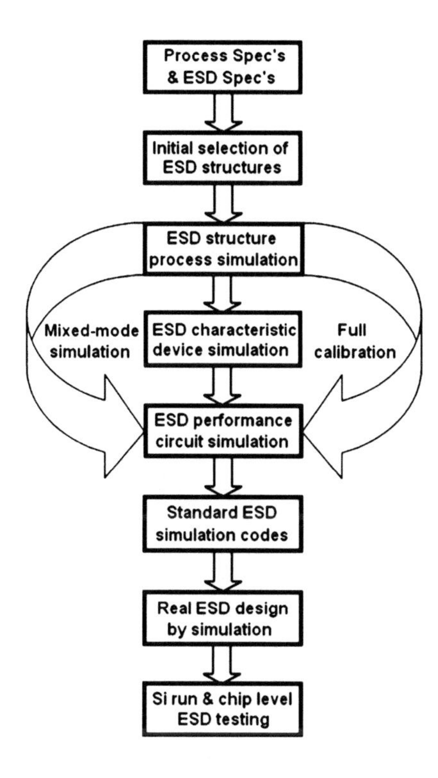

Figure 4-13. Representation of typical ESD development flow. (Adapted from [14].)

Moving from the given technology to the next one is a major challenge for ESD development, since in general there can be no simple and smooth transfer of protection concepts from the preceding technology. Traditionally, ESD development flow uses quantitative process data at an early stage [13, 14]. Based on the detailed description of the process steps, commercial process simulators can generate 2-D and 3-D structures, which include all the information about materials, topology and doping profiles. Using the simulated structures as an input for a device simulation, all essential ESD parameters can be examined. The standard ESD design methodology based on process and device simulations is shown in Figure 4-13. This methodology can be implemented in companies that have their own fabs for chip production, since they have the detailed information about the technology process [15]. However, the fabless companies are typically have limited technology information and they can not properly perform process simulations. In addition, they can not modify technology steps to optimize ESD devices, because ICs technology is typically optimized for functional or core transistors, but not for ESD devices. Hence, the ESD design methodology, shown in Figure 4-13, can not be practically used by fables companies. Therefore if the process simulations are not possible due to the limited technology information, the following strategy for ESD device calibration can be used.

- At the first step, all available information from the technology file in a given circuit simulation environment should be collected for the given technology. For example, in case of ESD MOSFET it can be gate oxide thickness (T_{ox}), effective channel length (L_{eff}), channel and poly-silicon dopings, threshold voltage (V_{th}), leakage and drive currents at normal operating conditions and device geometry parameters. These parameters should be used to create the device structure in a device simulator and calibrate it at normal operating conditions.
- For ESD devices, accurate high-current device models, which adequately capture the device breakdown and post-breakdown behavior, are of particular interest. For MOSFET devices this means that simulation models must match experimental triggering voltage (V_{t1}) and post-snapback "on" resistance (R_{on}). Mixed-mode simulation is used to run a snapback analysis of the device. Less well known device parameters such as typically doping profile are adjusted to achieve a match to available experimental data, which can be TLP *I-V* curve. This procedure is generally known as inverse modeling.
- If sufficient agreement could not be achieved by previous steps, the physical model coefficients in device simulator are adjusted as a final calibration step.

This calibration strategy was used for design of ESD protection devices (MOSFET and LVTSCR) implemented in 180 nm and 130 nm CMOS technologies [16, 17]. The synthesized MOSFET structure based on the reverse engineering approach is shown in Figure 4-14. The doping profile parameters were adjusted to match electrical test data, which included *I-V* characteristics at normal operating conditions and TLP data.

A mixed-mode (circuit-device) simulator was used to perform quasi-DC snapback simulations. The obtained results were compared with TLP data as shown in Figure 4-15. The TLP pulse duration was 100 ns and the rise time (10–90%) was 10 ns. The excellent agreement is seen between measured

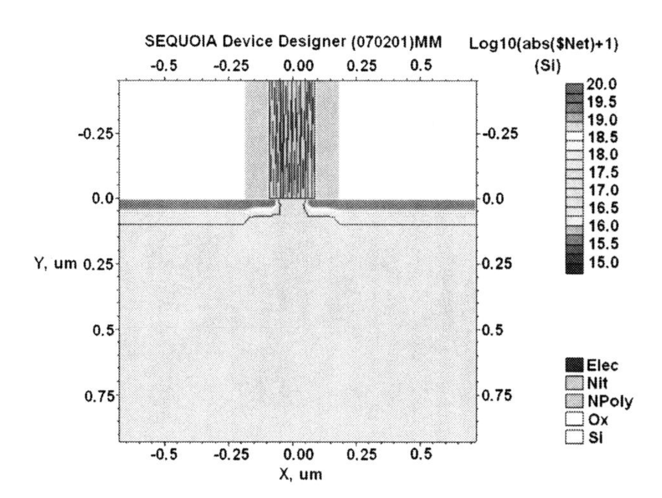

Figure 4-14. Synthesized ESD N-MOSFET structure (0.18 μm CMOS technology). (Adapted from [16].)

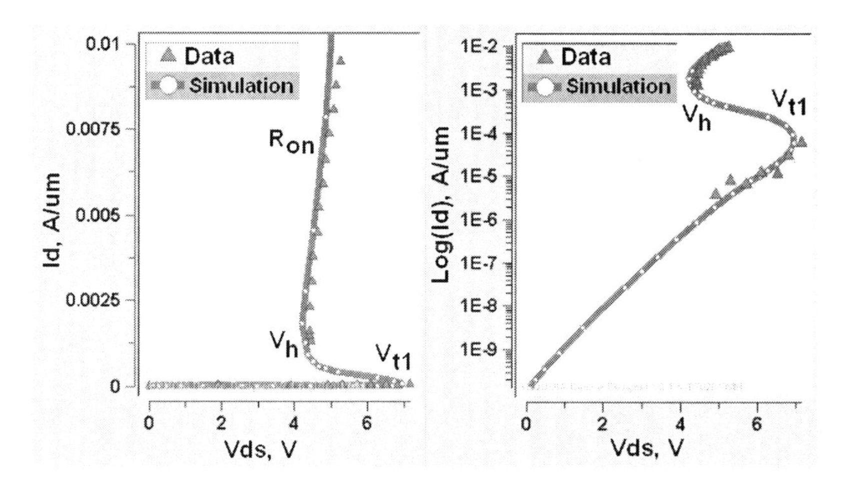

Figure 4-15. ESD N-MOSFET (50 μm width): I-V TLP data and simulation results. (Adapted from [16].)

data and simulation results. A good match was obtained for the key ESD parameters such as triggering voltage (V_{t1}), holding voltage (V_h) and post-snapback "on" resistance (R_{on}).

4.2 Mixed-Mode ESD Simulation

The mixed-mode (device-circuit) simulation provides the capability for in-depth studies of device level effects as well as analysis of larger circuits with complex interactions within of I/O buffer circuits embedded in chip environment. The mixed-mode simulations are widely used for the detailed analysis of ESD events in relatively small (10–20 devices) circuits [14, 17, 18]. Each active device in these circuits is implemented as a Finite Element

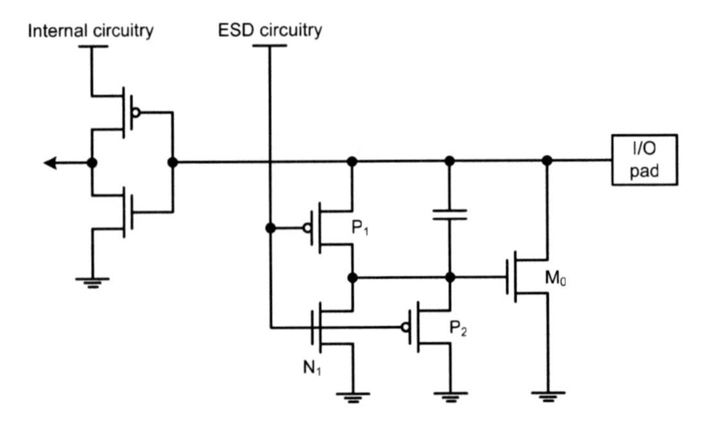

Figure 4-16. I/O ESD protection circuits simulated in mixed-mode. (Adapted from [18].)

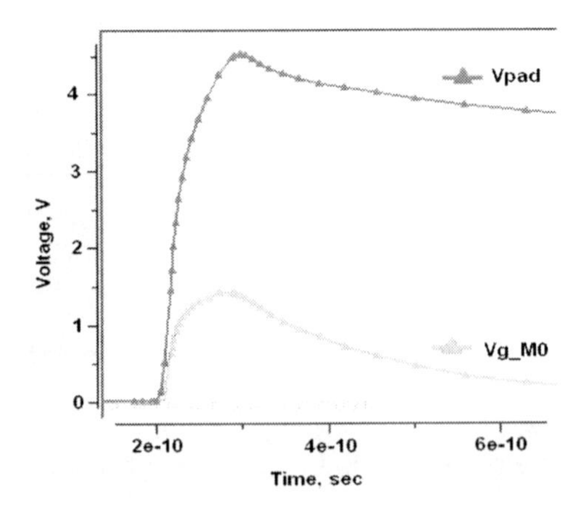

Figure 4-17. Mixed-mode simulation results of circuit shown in Figure 4-16 at 2 kV HBM ESD stress. (Adapted from [18].)

(physical structure) model (FEM) optimized and calibrated in a previous step of ESD design flow. The number of devices, which can be simultaneously simulated, is relatively small due to the significant increase of simulation time with increase of number of active devices. Figure 4-16 represents the ESD protection circuit and I/O pre-buffer simulated in mixed-mode. Each transistor in this circuit was implemented as a physical device structure (see Figure 4-14) optimized and calibrated for 180 nm CMOS technology.

The results of mixed-mode simulations at 2 kV HBM ESD stress are depicted in Figure 4-17 [18]. In this figure, Vg_M0 is the gate voltage of ESD transistor M0 and Vpad is the pad voltage at 2 kV HBM ESD stress applied to the I/O pad.

4.3 Chip-Level ESD Simulation

Mixed-level circuit-device simulation is a powerful tool for the ESD stress analysis of semiconductor products. Physical finite-element (FEM) level models are utilized for MOSFETs and other devices involved in the high-voltage high-current ESD event. Embedding these devices in realistic circuit simulation environment including probe, chip and interconnect parasitics (R, L, C) assures the required level of accuracy. CPU requirements (time and memory) for such simulations depend on the total number of degrees of freedom. Each circuit element contributes to this number, with FEM-level devices contributing the most. On a modern PC, complete I/O buffer circuits can be implemented with up to 20–30 MOSFETs modeled at the FEM level.

In certain cases it is possible to identify circuit blocks which are not involved in the high-voltage high-current ESD event. Therefore, simulating devices in these blocks on the physical FEM level is not necessary. Analytical models such as the industry standard BSIM3 model can be used instead to cover the low voltage range of operation. Some advanced ESD simulators, for example SEQUOIA ESD [19], have the capability to include a number of BSIM3 level devices along with FEM-level MOSFETs and other circuit elements (R, L, C). Automatic voltage/current checks can be included in low voltage circuit blocks to make sure that all BSIM3 devices operate in the allowed current/voltage range.

The ability to combine FEM-level and BSIM3 MOSFET models in one circuit greatly extends the applicability of mixed-mode analysis. Large circuits with substantial numbers (100 s) of low voltage device and a number of FEM-level devices can now be simulated on a regular PC. To make sure that all device models are operating in their valid range, automatic current/voltage monitors can be included in the circuit. If for example a peak voltage monitor indicates that a narrow core logic device is driven close to its triggering voltage, this indicates potential device failure. This simulation can then be repeated using a FEM-level model for the device in question to study its behavior in greater detail.

Let's consider an example where a mixed approach is possible with a part of the circuit described using BSIM3 MOSFET models while some circuit elements are subject to high-current high-voltage ESD stress and are modeled on the FEM level. We compare results produced by this mixed approach to a full FEM-level simulation. The example circuit is a gate-grounded MOSFET (ggMOSFET) based transient clamp with a low pass inverter circuit used to pull up the gate of the protection device for faster triggering and enhanced ESD protection [20]. The ggMOS ESD protection device is designed to enter impact-ionization induced snapback. It is therefore simulated on the physical FEM-level. The inverter circuit on the other hand is designed to provide a gate bias for the protection ggMOSFET and should not conduct a large current. Under certain conditions the inverter can be represented with BSIM3 MOSFET models.

To verify the validity of this hypothesis, two different simulations were performed using: (a) full FEM models for all MOSFETs, and (b) BSIM3 models for the inverter and an FEM model for the ESD MOSFET (M10) as shown in Figure 4-18. MOSFETs with a gate length of 0.25 μm and triggering voltage of about 9 V (FEM model, grounded gate) were used. In this figure, ESD MOSFET (M10) was implemented as a physical FEM model and transistors M11 and M12 in CMOS inverter were created as a BSIM3 models. Transient response of the transient ESD clamp circuit (Figure 4-18) during a Human Body Model (HBM) discharge is controlled by the RC time constant of the low pass filter C1, R6. If this time constant is properly selected, it raises the gate potential of the ESD MOSFET clamp long enough during the ESD pulse to help it absorb the ESD current. On the other hand, the RC time constant must be small enough to not negatively affect circuit performance. The choice of this time constant is therefore critical for adequate circuit operation. Figure 4-19 shows the peak pad voltage (VM2) at 2 kV HBM ESD stress as a function of (C1*R6) time for two cases, when all transistors were implemented as a FEM model and a mixed BSIM3-FEM model.

Both simulations show reasonably good agreement for larger time constants while higher discrepancy is observed for smaller time constants (C1*R6 ≤ 2 ns). Practically, traditional transient MOSFET based ESD protection circuits have the large RC network needed to trigger the clamp and keep it in a conducting state for the entire duration of the ESD event, which is ~600–700 ns for the HBM ESD stress. Hence, the mixed BSIM3-FEM model can be used for HBM ESD simulations for all practicalpurposes. Note that the BSIM3 compact model of transistors for SPICE-like simulations in TCAD environment can be extracted from the technology file provided by the fab or BSIM3 model can be extracted from the physical

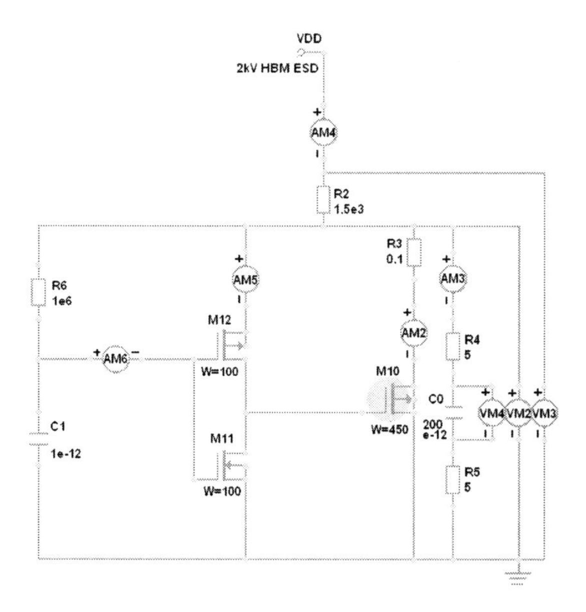

Figure 4-18. Transient ESD clamp circuit with FEM level transistor M10 and a BSIM3 level inverter (M11, M12). (Adapted from [20].)

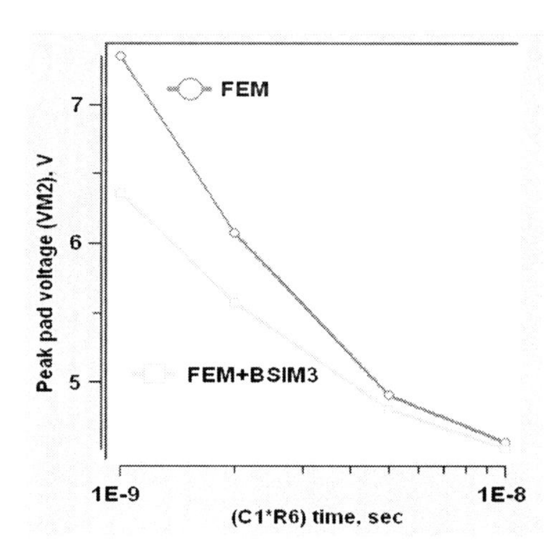

Figure 4-19. Peak pad voltage of transient ESD clamp circuit vs. (C1×R6) time calculated using the full FEM model and the mixed BSIM3-FEM model. (Adapted from [20].)

device model using automatic extraction tool embedded in some ESD/device simulators, for example Sequoia ESD [19].

Simulation techniques are widely used for the development of ESD protection on various levels reaching from process to circuit simulation. However, due to the complexity of the required simulation the ESD simulation of a complete IC is still a major challenge. Even under normal operational conditions a full chip compact simulation is not applicable to designs with 100 million gates or more. Beyond this, the ESD-relevant models for the compact simulation are even more complex, including the snapback behavior of ESD transistors and thermal effects. Fortunately, for the analysis of the overall voltage drop in the power net during the ESD discharge a very much simplified *I-V* characteristics can be used neglecting the snapback mode. It was shown that all ESD devices operating in a snapback mode can be descried by a bimodal *I-V* characteristic, where each branch is monotonic as shown in Figure 4-20 [4, 15]. One of the states reproduces the breakdown characteristic of the device without triggering the bipolar transistor, i.e. without the snapback, the second state describes the high current branch of the device after the snapback. Both characteristics consist of a high-ohmic regime modeled by a linear I-V dependency, which changes to a low-ohmic behavior (R_{diff} and R_{bd}) beyond the breakdown voltage (V_{bd}) and the holding voltage (V_h), respectively. In the standard circuit simulators, both states of the bimodal characteristic are examined and the worst case of ESD scenario is extracted, which is determined in accordance with the given physical constraints. This approach avoids the

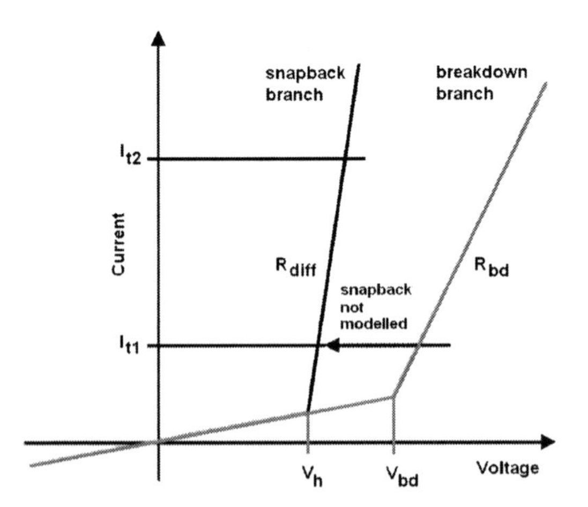

Figure 4-20. ESD devices and circuit are modeled either in the breakdown or snapback branch of their I-V characteristic. The snapback mode is not modeled.

necessity to consider transient effects. It can be reasonable for HBM and MM discharges, where the total current of the ESD event is a well-defined by the external parameter. A damage is detected, if the critical current I_{t2} or the upper limit voltage of the ESD design window is exceeded. This method was used to analyze the ESD robustness of a pad ring of a 300 pin IC. It was simulated for 1,600 different stress combinations on a 360 MHz RISC processor within 8 h [4].

In case of CDM stress, the ESD analysis is more complicated since the discharge current is determined by the charge stored on the IC. This requires the consideration of the capacitances of the supply networks and the package. A transient analysis including the RC times of the different discharge paths should be performed.

4.4 Test Chip Development

The robustness of ESD protection devices is very sensitive to layout. The thermal damages of ESD protection structures usually come from current crowding induced local over-heating stemming from layout discontinuities. For example, the current crowding occurs at the corners and edges of diffusion/metal layers. Therefore, layout is an extremely important for realizing robust ESD protection circuits. Figure 4-21 shows the layout examples of devices (single finger MOSFET on the top and multi-finger

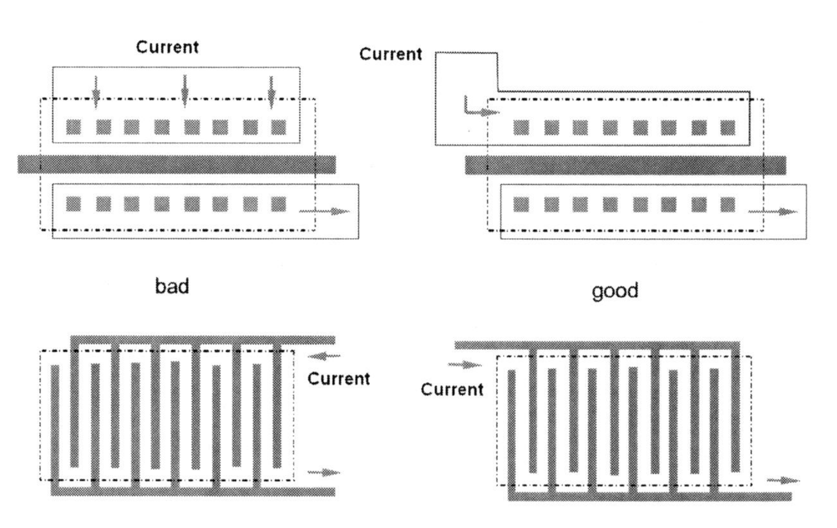

Figure 4-21. The layout of ESD MOSFETs to avoid current crowing: the single finger device shown on the top row and the multi-finger device shown on the bottom row. (Adapted from [1].)

MOSFET on the bottom) to avoid current crowing in the ESD device [1]. Unfortunately, layout issues of ESD protection devices can not be analyzed using 2-D device simulators, which are typically used in ESD protection circuit design flow. To investigate the impact of layout on ESD robustness, the ESD test structures should be implemented in silicon. The layout design guide for ESD protection devices was developed by SEMATECH [21].

Nowadays, the design of modern high performance digital and analog RF circuits leaves very small window for ESD circuits design due to the high sensitivity of the core of high speed digital and RF circuits to any parasitics. The most critical components are the low noise amplifier (LNA) and the power amplifier (PA) due to the high requirements on their RF performance. To avoid the RF circuit performance degradation, the careful minimization of the ESD device RC parasitics is required [22–24].

For example, a typical requirement on the total input parasitic capacitance (C_{par}) of a 2 GHz CMOS LNA is in the order of few hundreds fF. The actual limiting factors are the required LNA gain and noise figure. Both parameters are direct functions of C_{par} and degrade when C_{par} or the operating frequency increase. To estimate the impact of ESD parasitics on the performance degradation of high speed digital and analog RF circuits, the test circuits with ESD protection elements should be implemented in silicon. The LNA with ESD protection elements can be used to verify the ESD RC parasitics, as it shown in Figure 4-22. Note, that instead of shown ESD protection elements other ESD devices can be used for RC paracitics verification.

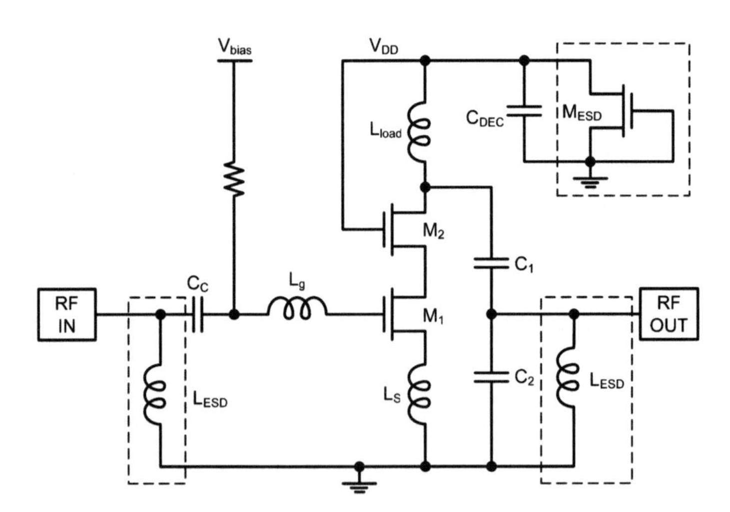

Figure 4-22. Common source LNA with added ESD protection within dotted boxes. (Adapted from [23].)

In order to examine the impact of ESD RC parasitics on high performance digital circuits, the test structure consisting of 40-stage ring oscillator (RO) was proposed [25]. The effect of ESD parasitics is determined by the degradation of operating frequency of ring oscillator due to the adding of ESD protection elements. The general concept of this test structure is depicted in Figure 4-23. For example, a 15-stage 4.7 GHz ring oscillator was implemented in 180 nm CMOS process [26]. In this test circuit the dual-directional NPNPN ESD structures [27] were used instead of ESD diodes. The measurement data shown, that in single-load case where only one node had ESD protection, a 50% clock speed reduction occurs due to the parasitic capacitances of ESD structures. It was found that optimized for 4 kV HBM stress dual-directional NPNPN ESD structure has 90 fF of parasitic junction capacitance and 29 fF of parasitic capacitance of Cu interconnects [26].

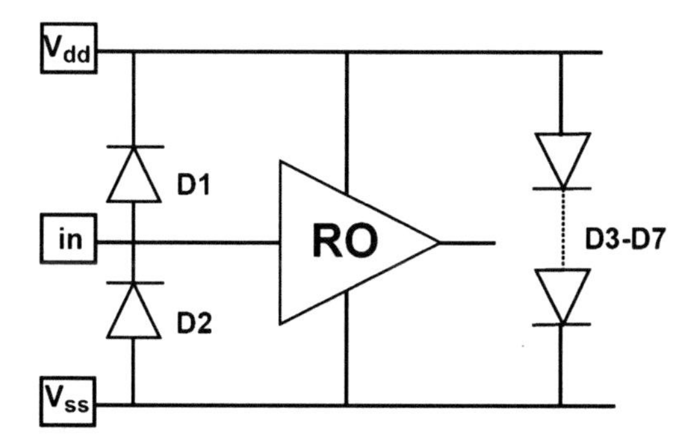

Figure 4-23. The test circuit diagram with p^+-nwell diode-based ESD protection network designed around the core (ring oscillator) circuit.

For mixed-signal ICs, the analysis of digital noise coupling to analog and design for its avoidance is very important. The sources of digital noise coupling are capacitive coupling, coupling through the power supply net, and coupling through the substrate. The noise is also generated through ESD protection networks in a multiple power supply system, which has different V_{DD} for digital and analog blocks.

Diode-based ESD power networks are commonly used in many applications due to easy implementation and area efficiency. However, these protection networks are not favored in view of noise margins needed for circuit operation. To verify the relationship between noise margin and ESD performance of diode-based power clamp circuits, the special test circuits was developed [5], as shown in Figure 4-24.

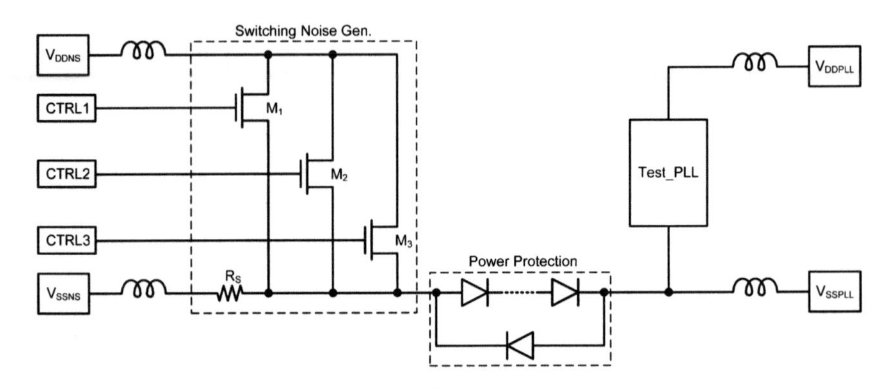

Figure 4-24. Test circuit diagram for measurement of PLL jitter under ground noise injected through the ESD power protection circuit. (Adapted from [5].)

Test structure consists of digital switching noise generator, an analog phase-locked loop (PLL) as noise monitor, and an ESD power clamp circuit as a Power/Ground noise propagation path. A surge of current from V_{SSNS} through a protection diode circuit inductively induces voltage fluctuations in V_{SSPLL}. A protection diode circuit forms a return path between two inductors on the V_{SSNS} and the V_{SSPLL} pins, which causes the effective potential of the PLL ground to fluctuate and thus changes the output frequency and phase of the voltage-controlled oscillator (VCO) integrated in PLL block. When V_{SSNS} is less than the turn-on voltage of the diode clamp, the jitter performance of PLL block has a slight influence on noise magnitude due to capacitive coupling. When the peak ground noise exceeds the turn-on voltage, the ground feed-through causes a significant increase of the jitter. From experimental results, it was found that the statistical distribution of jitter broadens as the ground noise becomes larger than the turn-on voltage of a clamp circuit. Thus, a higher number of diodes should be used for better noise margin [5]. However, the increasing the number of diodes raises the "on" resistance of a forward-biased diode clamp, which degrades the ESD performance.

4.5 ESD Measurements

The final step of circuit design flow for ESD protection is the ESD measurements. ESD tests determine the electrical properties of integrated circuits and their elements such as transistors, resistors and diodes at current levels and time scales typical for ESD events, amps of current and nano-seconds of time. The test method that has been almost universally adapted over the last few years is the transmission line pulse (TLP) measurement. The detailed description of TLP equipment and measurement technique was discussed in

Chapter 2. TLP measurements can be made on full circuits, in a way similar to HBM and MM measurements, they can be made on sub-circuits like input/output buffers or power supply clamps, or they can be made on individual circuit elements such as transistors or diodes [28]. Recent measurements of "real HBM" events indicate that under dry conditions the rise time is in fact much faster than the specifications for HBM testing [29]. For humid conditions the "real HBM" rise time becomes longer. It is therefore important that a protection scheme be able to function over a wide range of rise times, both to cover the wide range of rise times of the specification but also the wide range of real world rise times. TLP measurements can easily vary the rise time and verify protection over a range of rise times. HBM simulators are more limited. Due to equipment constraints, most high pin count simulators have rise times about 8–9 ns. The flexibility of TLP measurements allows the exploration of device performance over a wider range on stress conditions. TLP measurements can be made on sub-circuits, as already mentioned. As an example we consider a dynamically triggered power supply clamp, shown in Figure 4-25 [28].

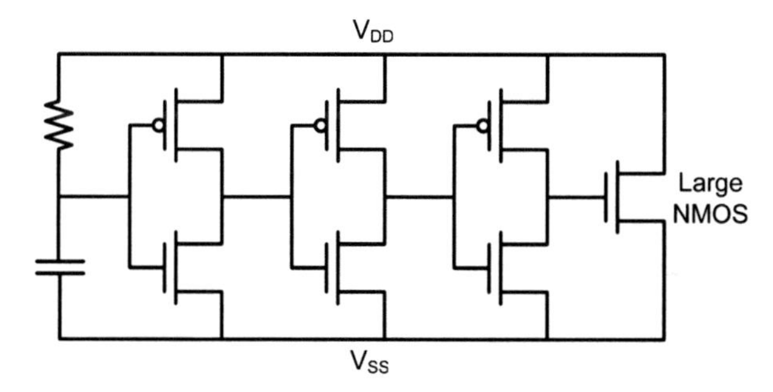

Figure 4-25. Schematic of dynamically triggered power supply clamp based on a large NMOSFET.

Power supply clamps are an important part of any ESD strategy. They are not only important for the protection of ESD stress between power supplies and ground, but can help protect inputs and output buffers with the use of steering diodes. Steering diodes are placed between Inputs/Outputs and power/ground buses and are used to direct an ESD stress through a power supply clamp. A TLP *I-V* measurement of a dynamically triggered power supply clamp is shown in Figure 4-26. At low current the device shows the expected linear behavior. At a little below 5 V, the I-V curve shows an increase in conductivity. This is likely the onset of bipolar action as the drain diode of large N-MOSFET reaches its avalanche point. This

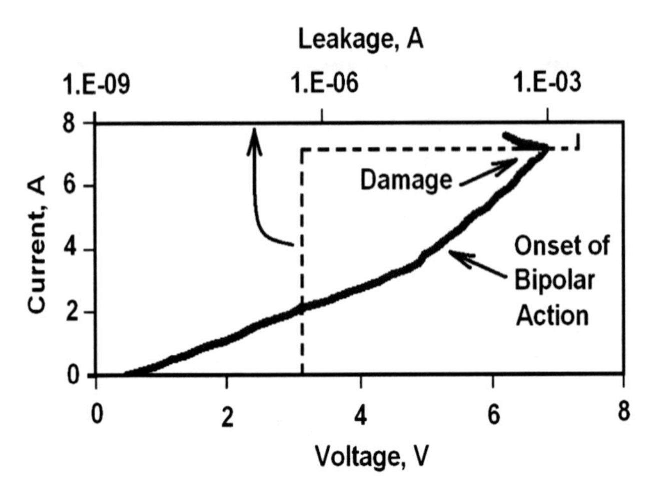

Figure 4-26. TLP I-V curve and leakage evaluation of power supply clamp. (Adapted from [28].)

provides an additional conductivity mechanism. At about 6.7 A the leakage increases dramatically and clamp is damaged by the high current.

Parameters extracted from the TLP measurements can be used to estimate the ESD behavior and ESD robustness of developed protection network. The obtained data can predict the triggering mechanisms such as bipolar snapback, calculate voltage drops for various current levels and predict the stress level at which device damage will occur. Generally, ESD test chip should contain not only the ESD protection elements themselves, but also the devices that constitute the output driver and the input receiver. Only when the *I-V* characteristics of each of these components are compared, the effectiveness of an ESD protection strategy can be determined.

The Human Body Model (HBM) and the Machine Model (MM) are the general ESD test methods, which are also widely used for commercial ICs. Generally, there are V_{DD}, V_{SS} and Pin-to-Pin ESD test modes, as shown in Figure 4-27.

Since ESD charge may have positive or negative polarity, each test mode should be performed twice. After I/O pins and power pins are tested by ESD simulator, ESD robustness of the IC is decided by the worst case. Most device manufacturers follow the general procedure defined by the MIL-STD-883C method 3,015:

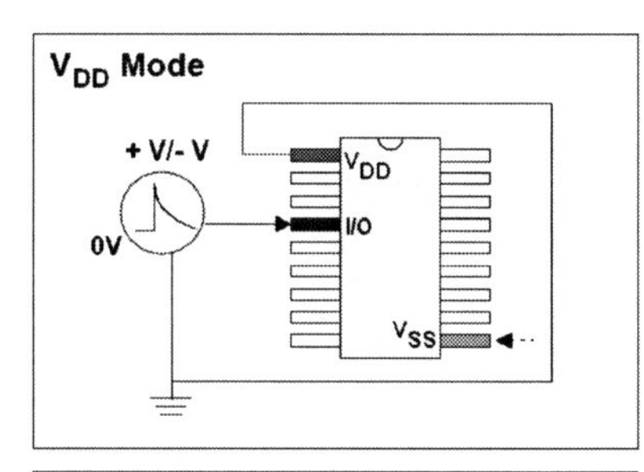

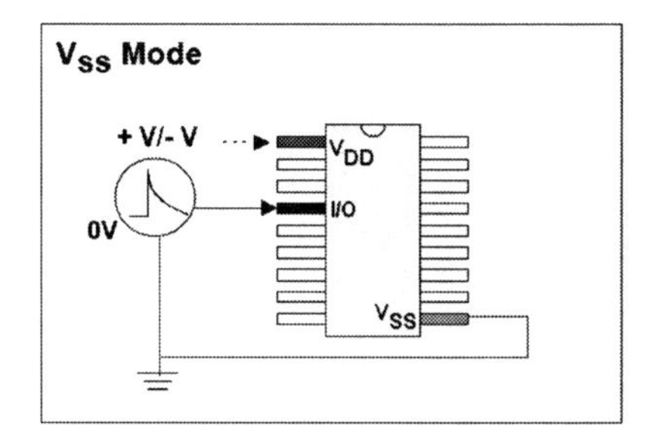

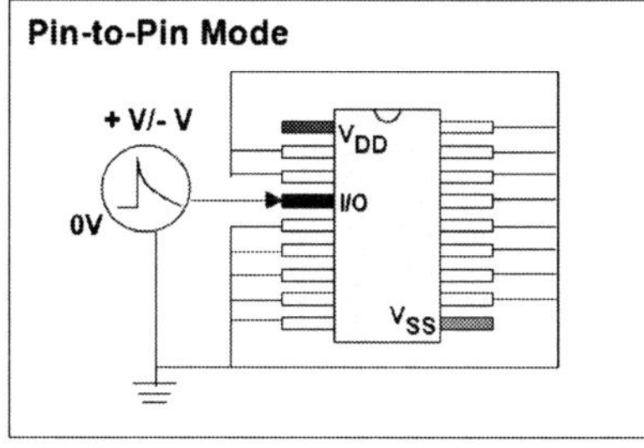

V_{DD} Mode :
V_{DD} pin is grounded.
Others pins (V_{SS} & I/O) are zapped, pin by pin

V_{SS} Mode :
V_{SS} pin is grounded.
Others pins (V_{DD} & I/O) are zapped, pin by pin

Pin-to-Pin Mode :
I/O pins are zapped, pin by pin.
I/O pins which are not zapped are grounded.
All power pins (V_{DD} & V_{SS}) are floated.

Figure 4-27. HBM and MM ESD test modes.

- Selection of Pin Combination: V_{SS} Mode, V_{DD} Mode, Pin-to-Pin Mode.
  - In principle, all the pins are required to be stressed against every other pin in turn, which is practically unfeasible.
  - Thus a reasonable judgment must be made for the worst case of pin-to-pin combination.
- Injection Pulses: Three positive and then three negative stress pulses with pulse interval at least a 300 ms between consecutive zaps.
- ESD Pass Level: Worst Case ESD Performance.
- Test Sample Number: At least three samples in each Modes & Stress Level.

The DC *I-V* curve after ESD zapping of each I/O, V_{DD} and V_{SS} pins is compared with the DC *I-V* curve before ESD zapping, to fix any curve shifting, which indicates that the ESD protection circuit is failed at the given ESD stress level.

5. SUMMARY

A good on-chip ESD protection network should have the following characteristics: (1) fast triggering to avoid premature ESD failure due to accidental turn-on of competing internal parasitic structure; (2) high current handling capability, good heat dissipation capability, and low discharging impedance to boost the ESD robustness; and (3) low parasitic effects to minimize negative impacts on core circuits. All these ESD features are critical for ESD protection design evaluation. ESD phenomena include different coupling effects, such as thermal, process, device, circuit, and layout issues. Starting from the sub-100 nm CMOS technologies, the protection development goes much beyond the development of a specific optimized protection element. A sophisticated protection network has to be designed, which covers both the I/O circuit and the core region, where low oxides thickness and low junction breakdown voltages lead to hard constraints on the maximum voltage overshoot during ESD. In especially, designs with multiple power supply domains will complicate the ESD supply protection concept extremely. To achieve a good ESD robustness it will be necessary to consider the ESD protection as integral part of the IC development starting from the concept phase. To support this and to extract the necessary data for an ESD optimization, the following steps of design flow of ESD protection network should be performed:

- Process/Device simulation and calibration
- Mixed-Mode ESD simulation
- Chip-level ESD simulation
- Test chip development
- ESD measurements

Today, due to the narrow design window and the higher complexity of the power supply concepts, the ESD protection becomes much more an overall problem of the IC, which has to be considered right from the IC's concept phase. As the technology continue to progress deep into the sub-100 nm range, process advances will be sure to have additional impact on ESD. With the development of good ESD modeling tools these problems should be predictable as well as solvable at the process development step. The ESD requirements thus need to be a part of the technology definition from the start.

REFERENCES

[1] J.-B. Huang and G. Wang, "ESD protection design for advanced CMOS," *Proc. of SPIE*, vol. 4600, pp. 123–131, 2001.

[2] R. Merrill and E. Issaq, "ESD design methodology," *EOS/ESD Symposium*, pp. 233–237, 1993.

[3] J. E. Vinson and J. J. Liou, "Electrostatic discharge in semiconductor devices: protection techniques," *Proc. of the IEEE*, vol. 88, No. 12, pp. 1878–1900, 2000.

[4] H. Gossner, "ESD protection for the deep sub micron regime – a challenge for design methodology," *Proc. of the Int. Conf. on VLSI Design (VLSID)*, pp. 809–818, 2004.

[5] L. Lee, Y. Huh, P. Bendix, S.-M. Kang, "Understanding and addressing the noise induced by electrostatic discharge in multiple power supply systems," *Proc. of the Int. Conf. on Computer Design (ICCD)*, pp. 406–411, 2001.

[6] M.-D. Ker and H.-H. Chang, "Whole-chip ESD protection strategy for CMOS IC's with multiple mixed-voltage power pins," *Proc. of the Int. Symp. on VLSI Tech., Systems, and Applications*, pp. 298–301, 1999.

[7] M.-D. Ker and H. C. Jiang, "Whole-chip ESD protection strategy for CMOS integrated circuits in nanotechnology," *Proc. of the IEEE Conf. on Nanotechnology*, pp. 325–330, 2001.

[8] M.-D. Ker, H.-H. Chang, and T.-Y. Chen, "ESD buses for whole-chip ESD protection," *Proc. of the Int. Symp. on Circuits and Systems (ISCAS)*, pp. 545–548, 1999.

[9] J. W. Miller, G. B. Hall, A. Krasin, M. Stockinger, M. D. Akers, and V. G. Kamat, "Electrostatic discharge protection circuitry and method of operation," *US Patent Application*, No. 2004/0027742A1, 2004.

[10] C. A. Torres, J. W. Miller, M. Stockinger, M. D. Akers, M. G. Khazhinsky, and J. C. Weldon, "Modular, portable, and easy simulated ESD protection networks for advanced CMOS technologies," *Microelectronics Reliability*, vol. 42, No. 6, pp. 873–885, 2002.

[11] M. Stockinger and J. W. Miller, "Electrostatic discharge protection circuit and method of operation," *US Patent Application*, No. 2005/007819A1, 2005.

[12] M. Stockinger, J. W. Miller, M. G. Khazhinsky, C. A. Torres, J. C. Weldon, B. D. Preble, M. J. Bayer, M. Afkers, and V. G. Kamat, "Advanced rail clamp networks for ESD protection," *Microelectronics Reliability*, vol. 45, No. 2, pp. 211–222, 2005.

[13] K. Esmark, W. Stadler, M. Wendel, H. Gobner, X. Guggenmos, and W. Fichner, "Advanced 2D/3D ESD device simulation – a powerful tool used in a pre-Si phase," *EOS/ESD Symposium*, pp. 420–429, 2000.

[14] H. Feng, G. Chen, R. Zhan, Q. Wu, X. Guan, H. Xie, A. Z. H. Wang, and R. Gafiteanu, "A mixed-mode ESD protection circuit simulation-design methodology," *IEEE J. of Solid-State Cir.*, vol. 38, No. 6, pp. 995–1006, 2003.

[15] K. Esmark, H. Gossner, and W. Stadler, *Advanced Simulation Methods for ESD Protection Development*, Elsevier, Oxford, 2003.

[16] SEQUOIA Design Systems Inc., *SEQUOIA News*, vol. 4, No. 1, 2005. http://www.sequoiadesignsystems.com/

[17] O. Semenov, H. Sarbishaei, V. Axelrad, and M. Sachdev, "Novel gate and substrate triggering techniques for deep sub-micron ESD protection devices," *Microelectronics Journal*, vol. 37, No. 6, pp. 526–533, 2006.

[18] K. Iniewski, V. Axelrad, A. Shibkov, A. Balasinski, S. Magierowski, R. Dlugosz, and A. Dabrowski, "3.125 Gb/s power efficient line driver with 2-level pre-emphasis and 2 kV HBM ESD protection," *Proc. of the Int. Symp. on Circuits and Systems (ISCAS)*, vol. 5, pp. 5071–5074, 2005.

[19] SEQUOIA Design Systems Inc., http://www.sequoiadesignsystems.com/

[20] SEQUOIA Design Systems Inc., *SEQUOIA News*, vol. 1, No. 2, 2002. http://www.sequoiadesignsystems.com/

[21] SEMATECH, "Test structures for benchmarking the electrostatic discharge (ESD) robustness of CMOS technologies," *Technology Transfer 98013452A-TR*, 1998.

[22] V. Vassilev, S. Thijs, P. L. Segura, P. Wambacq, P. Leroux, G. Groeseneken, M. I. Natarajan, H. E. Maes, and M. Steyaert, "ESD-RF co-design methodology for the state of the art RF-CMOS blocks," *Microelectronics Reliability*, vol. 45, No. 2, pp. 255–268, 2005.

[23] D. Linten, S. Thijs, M. I. Natarajan, P. Wamback, W. Jeamsaksiri, J. Ramos, A. Mercha, S. Jenei, S. Donnay, and S. Decoutere, "A 5-GHz fully integrated ESD-protected low-noise amplifier in 90-nm RF CMOS," *IEEE J. of Solid-State Cir.*, vol. 40, No. 7, pp. 1434–1442, 2005.

[24] P. Leroux, V. Vassilev, M. Steyaert, and H. Maes, "High ESD performance, low power CMOS LNA for GPS applications," *Journal of Electrostatics*, vol. 59, No. 3–4, pp. 179–192, 2003.

[25] N. Maene, J. Vandenbroeck, and K. Allert, "On-chip electrostatic discharge protections in advanced CMOS technologies," *Microelectronics Reliability*, vol. 32, No. 11, pp. 1545–1550, 1992.

[26] A. Wang, "Recent developments in ESD protection for RF ICs," *Design of Automation Conf.*, pp. 171–178, 2003.

[27] A. Wang and C.-H. Tsay, "On a dual-polarity on-chip electrostatic discharge protection structure," *IEEE Trans. on Electron Devices*, vol. 48, No. 5, pp. 978–984, 2001.

[28] R. A. Ashton, "Transmission line pulse measurements: a tool for developing ESD robust integrated circuits," *Proc. of the Int. Conf. on Microelectronic Test Structures*, vol. 17, pp. 1–6, 2004.

[29] J. Barth, J. Richner, L. Henry, and M. Kelly, "Real HBM & MM – The dV/dt threat," *EOS/ESD Symposium*, pp. 179–187, 2003.

Chapter 5

ESD POWER CLAMPS

1. INTRODUCTION

Whole-chip ESD protection has become an important reliability issue of CMOS IC's. Even if there are suitable ESD protection circuits at the input and output pads, the internal circuits are still vulnerable to the ESD damage. The pin-to-pin ESD stress, as shown in Figure 5-1, often causes some unexpected ESD damage located in the internal circuits, rather than the input or output ESD protection circuits. In Figure 5-1, the ESD current is discharged through the internal circuit and may cause random ESD damage in this circuit, as the "Path 1" current flow shown in Figure 5-1. If there is an effective ESD clamp circuit across the V_{DD} and V_{SS} power buses, the ESD current can be discharged through the "Path 2" current flow. Thus, an effective ESD clamp circuit between the power rails is necessary for the protectting of internal circuits against the ESD damage. Clamps can come in many different varieties. A simple diode clamps were already mentioned in Chapter 4 (Figure 4-2).

ESD Clamps can be grouped into two categories: static and transient. The static clamps provide a static or steady-state current and voltage response. A fixed voltage level activates static clamps. As long as the voltage is above this level, the clamp will conduct current. A diode, MOSFET and SCR based clamps are known as static ESD clamps. Transient clamps take advantage of the rapid changes in voltage and/or current that accompanies an ESD event. During this transient, an element is turned on very quickly and slowly turns off. This type of clamp conducts for a fixed time when it is triggered. An RC

O. Semenov et al., ESD Protection Device and Circuit Design for Advanced CMOS Technologies, 117–146.

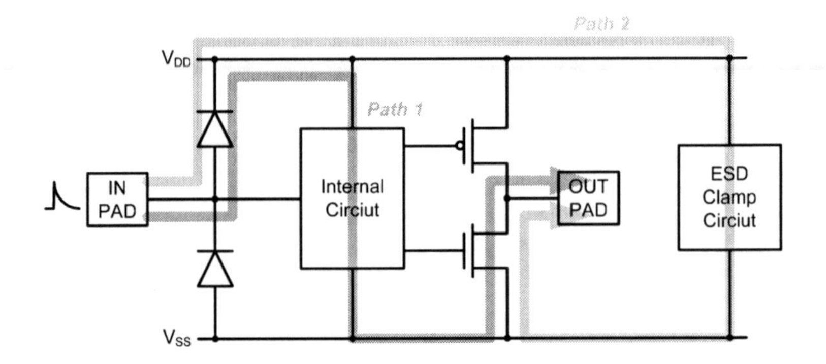

Figure 5-1. The ESD current discharging path in a IC during the pin-to-pin ESD stress event.

network determines the time constant. These clamps are typically triggered by very fast events on the supply lines.

Both groups of clamps have their respective advantages and disadvantages. The static clamp typically occupies less space and is composed of fewer elements. If a static clamp falsely triggers while power is applied to the part, it will result in large, steady current and may lead to damage. Transient clamps, on the other hand, can be designed to turn on very quickly and handle larger transient events. The disadvantage is that they will also respond to any fast event, even noise. If they falsely trigger while the part is powered, they could interfere with circuit operation and it is likely that the part will be destroyed. In this chapter, the different designs of ESD clamps are considering and their advantages and disadvantages are discussed in details.

2. STATIC ESD CLAMP

The supply clamp should limit the supply voltage by conducting a large amount of current through a low impedance path when the trigger voltage is exceeded. The trigger voltage is generally set higher than the maximum supply voltage to minimize current during normal operation. The selection of what type of clamp to employ in a design is based on several criteria. The first criterion is what circuit elements are available in the given process. The next is related to the environment (V_{DD}, temperature) where the circuit must operate. The other criteria that affect the decision are its current handling capability and its turn-on time. Turn-on time is especially important for CDM ESD stress since this type of ESD is a very fast event. Many clamps may not be able to respond quickly enough to provide protection against this type of ESD event.

2.1 Diode-Based ESD Clamps

Because the diode in the forward-biased condition can sustain a much higher ESD level than it in the reverse-biased condition, the diode string with multiple stacked diodes was therefore proposed to clamp the ESD overstress voltage on the 3.3–5.0 V tolerant I/O pad [1] or between the mixed-voltage power lines [2, 3]. The diode based ESD power clamps can be developed by using Zener diode [4], diode strings [5] or cantilever diode strings [2, 6], as shown in Figure 5-2. During the ESD event, the V_{DD} pad voltage rises above

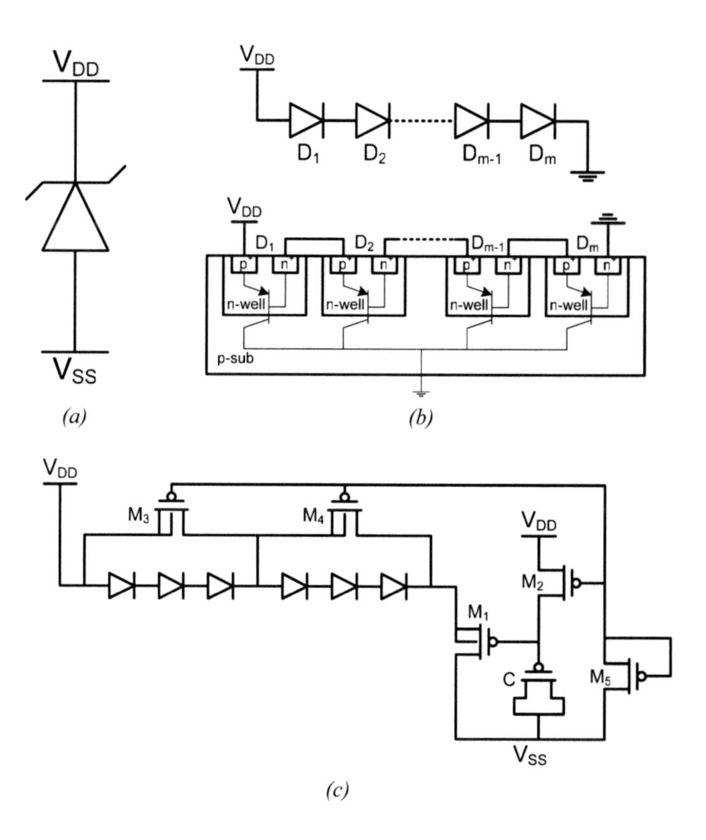

Figure 5-2. Diode based ESD clamps: (a) Zener diode, (b) diode string and its cross-sectional view, (c) cantilever diode string.

the Zener breakdown voltage causing the dissipating of ESD current through the Zener diode (Figure 5-2(a)). This protection method requires a Zener breakdown voltage above the V_{DD} voltage. The implementation of Zener diodes and their operating principals are explained in Chapter 3. The disadvantages of ESD clamps based on Zener diodes are that this device is not readily available in many CMOS process and typically requires additional process steps (silicide block option). In addition, the breakdown voltage of Zener diodes is a process dependent since it's determined by the accurate control of doping densities in the diffusion region.

The diode string shunts the ESD current to ground through the parasitic vertical *pnp* bipolar transistors that are formed in silicon substrate, as shown in Figure 5-2(b). In the ideal case, the triggering (turn-on) voltage of a diode string can be linearly increased when more diodes are stacked in the diode string. However, if the gain of parasitic vertical *pnp* transistor (β) is one or even larger, the addition of stacked diodes in the diode string doesn't linearly increase the triggering voltage across the diode string (see equation (5-1) [5]), but causes more leakage current flowing into the substrate.

$$V_{\text{trig.}}(I) = mV_D(I) - nV_T \times \left[\frac{m(m-1)}{2}\right] \times \ln(\beta + 1) \qquad (5\text{-}1)$$

where $V_D(I)$ is the forward turn-on voltage of one diode, I is the current flowing into the diode string, V_T is the thermal voltage, n is the ideality factor, m is the number of stacked diodes in the diode string, β is the beta gain of the parasitic vertical *pnp* transistor.

This means that more stacked diodes would be needed to support the same triggering voltage of a diode string at a specified current when the gain of the parasitic vertical *pnp* increases. The main issue of the *pnp*-based diode string used as the V_{DD}-to-V_{SS} ESD clamp circuit is the leakage current, where the amplification effect of multistage Darlington beta gain often causes more leakage current from V_{DD} to the grounded substrate. If the voltage difference between V_{DD} and V_{SS} becomes larger, the leakage current will increase exponentially.

Another issue on the diode string is the reduction on the triggering voltage of the diode string at a higher temperature, as depicted in Figure 5-3. The typical forward biased turn-on voltage of diode is approximately 0.6–0.7 V at room temperature. At high (burn-in) temperature (110–125°C), the diode string leakage current is increased due to the Darlington effect and the diode turn-on voltage is reduced from 0.6 to 0.4 V. Generally, the turn-on voltage of *pn* junction decreases by approximately 20% every 100°C of

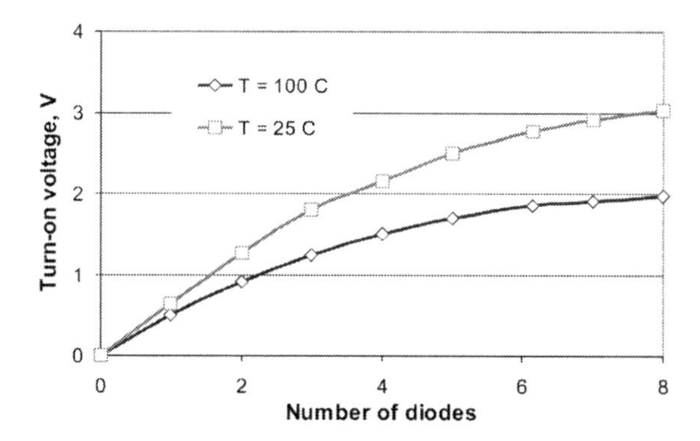

Figure 5-3. Diode string turn-on voltage for two temperatures (β = 6). (Adapted from [9].)

temperature increase [7]. As a result, a longer diode string will be needed to provide the same EOS/ESD protection at stress temperature and to prevent the false triggering under voltage spikes in burn-in oven. However, the increase in number of diodes and their series resistance may have effect on ESD reliability [8]. Hence, the ESD clamps based on diode string are not recommended for applications, which require the burn-in qualification and testing.

The design concept of the cantilever diode string is to block the diode string from V_{SS} when the IC is in the normal operating mode. It helps to avoid the amplification effect of multistage Darlington beta gain in the vertical *pnp* transistors formed in the diode string and hence to reduce the leakage current into the substrate. In Figure 5-2(c), the PMOS transistor M1 is used as the termination of the diode string from V_{SS} in normal operating mode. But it sinks a substantial amount of current while an ESD pulse occurs. The PMOS transistor M2 and MOS-capacitor C are used as a RC-based ESD detection circuit to distinguish the ESD stress from the normal V_{DD} voltage and to turn on or off M1 correctly. The PMOS transistors M3 and M4 are long channel devices which are used as the bias network. These transistors provide forward current from V_{DD} to the later staked diodes. The small NMOS transistor M5 provides a ground connection without allowing a power supply voltage drops across a single thin gate oxide. The device dimensions (W/L) of M1, M2, M3, M4 and M5 were optimized as 200/1, 1.8/40, 1.8/40, 1.8/40 and 1.8/5 (μm/μm), respectively, in a 0.35 μm silicide CMOS process [2]. Cantilever diode strings were developed to reduce the leakage current in diode based ESD clamps operating in the high-temperature environment.

2.2 MOSFET-Based ESD Clamps

The grounded gate ESD structure, shown in Figure 5-4(a) typically consists of a large n-channel grounded gate MOS transistor (GG-NMOSFET). The width of the transistor is large, typically 800 μm [10], to allow for a quick discharge of the ESD current without damaging the ESD device. This ESD structure is placed such that the drain is connected to V_{DD} and source together with gate being connected to V_{SS}. The advantage of this structure is the relatively small size which occupies a significantly lower silicon space. The triggering time for this clamp however sometimes can be a problem if there are other internal circuitry transistors which have shorter drain to gate spacing compared to the grounded gate ESD structure. These internal circuitry transistors might trigger faster than the grounded gate ESD structure and be damaged during the ESD event. Reducing the drain to gate spacing of the ESD structure may pose problems as the poor ballasting effect due to a shorter gate to drain spacing. Since GGMOSFET has a relatively big size, it should be implemented as a multi-finger device. The design of GG-NMOSFET ESD clamp must provide the simultaneous triggering of all fingers. Otherwise, GGMOSFET is typically failed at relatively low ESD stress [11].

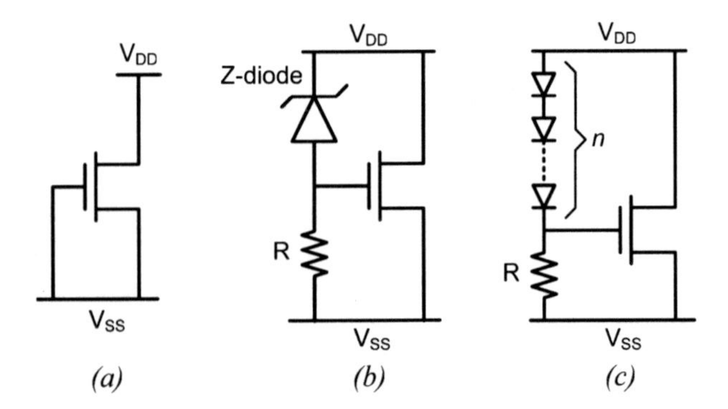

Figure 5-4. MOSFET based static ESD clamps: (a) Gate Grounded N-MOSFET (GG-NMOSFET), (b) Zener diode coupled N-MOSFET (ZCNMOS), (c) polysilicon diode string coupled N-MOSFET.

To reduce the turn-on voltage of GG-NMOSFET, ESD clamps with gate triggering by the Zener diode and stacked polysilicon diodes were developed, as depicted in Figure 5-4(b, c) [11, 12]. The advantage of Zener diode coupled NMOS ESD clamp (ZCNMOS) is the early triggering in comparison with the GG-NMOSFET ESD clamp. The ESD clamp with diode string has the stacked polysilicon diodes, which are operating in forward-biased condition during the ESD stress. The gate-triggering mechanism of NMOS is operated by using DC voltage level detection. Therefore, it is not degraded by the parasitic power line capacitance (C_{VDD}). If the stacked diodes are realized by the p^+ diffusion in n-well, as shown in Figure 5-2(b), the diode string has a significantly leakage current in the order of mA from V_{DD} to V_{SS} due to the parasitic vertical BJT effect in CMOS process [5]. But, when the stacked diodes are realized by the polysilicon layer, the leakage current can be reduced below 1 μA under 5 V V_{DD} bias. To minimize the leakage current in ESD clamp at normal operating conditions, the optimized values of resistor R and number of diodes *n* in the string are 100 kΩ and from 6 to 9 diodes, respectively [12].

2.3 SCR-Based ESD Clamps

Due to the inherent capability of high power delivery, the lateral SCR device has been used as an on-chip ESD protection element in CMOS ICs. Experimental results had shown that the lateral SCR device can sustain high ESD stress within a smallest layout area as compared to other traditional ESD protection elements [13]. An improved design on the lateral SCR is to use the low voltage triggered lateral SCR (LVTSCR) device to protect CMOS ICs. Due to the low trigger voltage (~10 V in 0.35 μm silicide CMOS process), the LVTSCR device can perform excellent on-chip ESD protection without the support of the secondary protection circuit. But, its low holding voltage/current may cause the LVTSCR to be accidentally triggered on by the external noise pulses while the CMOS IC is in the normal operating

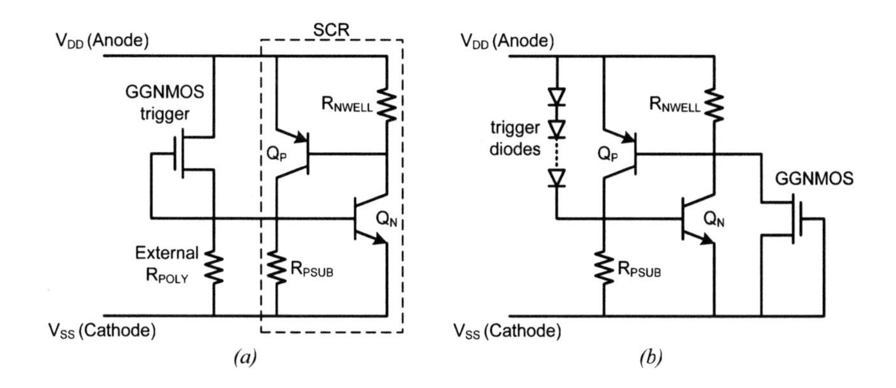

Figure 5-5. The general schematics of the high holding current LVTSCR (a) and the high holding voltage LVTSCR (b) optimized for ESD power clamps.

condition [14]. To solve this problem the LVTSCR based ESD clamps with high holding current and high holding voltage were developed, as depicted in Figure 5-5(a) and Figure 5-5(b), respectively.

In the high holding current LVTSCR clamp [15], R_{NWELL} and R_{PWELL} resistance values can be significantly reduced by adjusting the external shunt resistor and using effective SCR p-well/n-well layout techniques. Thus, both parasitic BJTs (forming the SCR) require more base current to reach the forward bias of the corresponding emitter-base junctions, which results in the regenerative SCR latch mode. Consequently, a higher triggering current (I_{t1}) as well as holding current (I_h) are required to trigger and to sustain the SCR on-state, respectively. The external polysilicon resistor value is approximately 10 Ω. The high-I_h LVTSCR clamp has the holding current by 3 × higher than the conventional gate grounded LVTSCR ESD clamp with the same size.

In the high holding voltage LVTSCR clamp [16–18], the diode chain is used to trigger LVTSCR during ESD stress conditions. The number of trigger diodes (*n*) must be chosen sufficiently high such that the chain does neither leak nor trigger the SCR during normal operating conditions. Conversely, the LVTSCR ESD trigger voltage is increasing with *n*. A reasonable design trade-off between low leakage at normal operation conditions ("maximize *n*") and low ESD trigger voltage ("minimize *n*") at ESD event can be achieved in the wide range of voltage applications up to 5.5 V. To use the diode chain triggered LVTSCR device (DTLVTSCR) as a power-clamp, its holding voltage must be increased above V_{DD} for latch-up immunity. Typically, two or three diodes are used in DTLVTSR.

3. TRANSIENT POWER CLAMPS

Transient clamps take advantage of the rapid changes in voltage and/or current that accompanies an ESD event. During this transient, an element is turned on very quickly and slowly turns off. This type of clamp conducts for a fixed time when it is triggered. An RC network determines the time constant. These clamps are typically triggered by very fast events on the supply lines. Several key advantages of transient power clamps include its ability to provide ESD protection at low trigger voltages, no added process steps, relaxed layout constrains, and circuit simulators such as SPICE can be utilized to design transient clamps [19]. The main disadvantage is that they will also respond to any fast event, even power supply noise or fast supply ramp up. If they falsely trigger, while the part is powered, they could interfere with circuit operation and it is likely that the part will be destroyed.

3.1 MOSFET and SCR-Based Transient Clamps

The schematics of typical MOSFET and SCR based transient clamps are shown in Figure 5-6. They can provide an additional shunt path for the current during an ESD event. These transient clamps make use of the rapid dv/dt of an ESD pulse to capacitively couple charge onto an internal control node [20]. The charge or voltage on this node is used to trigger active circuitry, such as a transistor, that provides a discharge path for the ESD pulse from the positive rail to the negative rail. Eventually, the control node is discharged through resistor R and the active clamp circuitry is shut off. These approaches have been used successfully on many designs [5, 11, 21]. As mentioned before, the disadvantages of these approaches are limited noise immunity and the potential of clamping during power supply turn-on. To design suitable RC-based circuit, the RC time constant must be designed about 1 microsecond to trigger the clamp and keep it in a conducting state for the entire duration of the ESD event which is approximately 600–750 ns for the HBM ESD stress. To satisfy this specification, R is typically realized with a 50 kΩ n-well resistance and C is realized with a 20 pF poly-gate of NMOS transistor. Large resistance and capacitance needed for the time constant occupies a large silicon area. One should carefully simulate maximum supply (C_{VDD}) and decoupling capacitance as these capacitances are likely to affect the triggering efficiency of the transient clamp. This is because the ESD event should charge not only the capacitor C in the control circuit, but the parasitic capacitor C_{VDD} as well (see Figure 5-6(a, b)). As a result, the turn-on time of ESD clamp is increased.

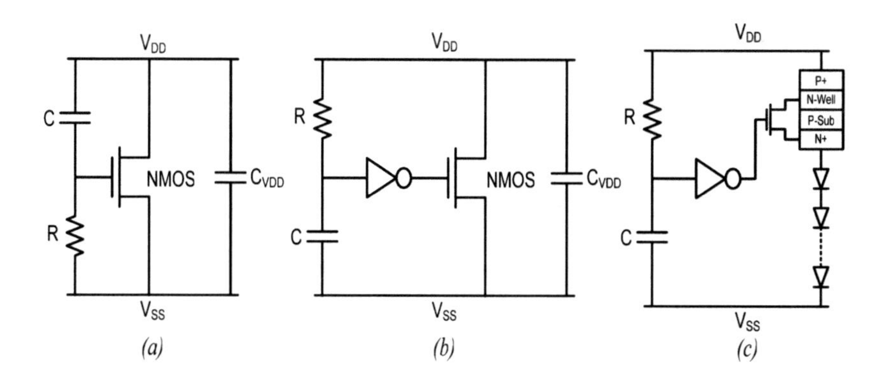

Figure 5-6. Transient MOSFET and SCR based ESD clamps: (a) Gate coupled NMOSFET, (b) RC control ESD clamp, (c) LVTSCR based ESD clamp with diode string.

Key design parameters in the supply clamp are the trigger voltage, the delay in turning on the clamp and the duration of the clamp. Generally, ESD clamp designs can not be ported from one process to another because of the changes in devices characteristics and breakdown voltages from process to process. However, a general approach can be employed and verified through circuit and device simulations. The clamp should limit the power supply to a voltage less than the breakdown voltage of any device and the clamp duration must last until the ESD event has decayed below any device breakdown voltage. The size of the clamp device is determined by the voltage allowed, characteristics of the clamp, bus resistance, and the voltage drops in cumulative series resistance including that of diodes in the ESD path.

The power-rail ESD clamp shown in Figure 5-6(c) [5] has high ESD robustness. It can avoid disadvantages of both the leakage current in the conventional diode string designs and the latch-up problem in the SCR device from V_{DD} to V_{SS}. Because the holding voltage of the LVTSCR in bulk CMOS technologies is only 1–1.5 V, the core circuit can be destroyed if the LVTSCR in the ESD protection circuits is accidentally turned on when the IC is in the normal operating condition. Therefore, the LVTSCR or SCR with a holding voltage lower than V_{DD} should be suitably modified to overcome the latch-up issue before they are used in the on-chip ESD protection circuits. To overcome the latch-up danger, the holding voltage of this new diode string should be designed greater than the voltage difference across the power rails by adding the more stacked diodes in the diode string. An LVTSCR device is added on the top of the diode string to block the leakage current from V_{DD} to V_{SS} through the diode string. But, the LVTSCR is designed to be quickly turned on to bypass the ESD current through the diode string, while the IC is in the ESD-stress condition. To achieve such desired turn-on and turn-off operations on the LVTSCR, the RC-based ESD-detection circuit is applied to control the gate of the LVTSCR. The RC time constant in the RC-based ESD-detection circuit is designed about 0.1–1 μs to distinguish the ESD-stress transition or the normal V_{DD} power-on transition. The HBM ESD voltage has a rise time about 10 ns. Under the normal V_{DD} power-on condition, the V_{DD} power-on waveform has a rise time in the range of milliseconds.

3.2 Three-Stage Transient Power ESD Clamp

Noise immunity during normal operation is an important design parameter for transient power clamp designs. From power on to power down, the power supply is subject to numerous sources of noise. The RC timer is responsible for distinguishing between normal noisy power events and an

ESD event. If the RC timer is incapable of making a successful differenttiation, the transient power clamp structure will likely to cause false triggering. In order to provide improved noise immunity, two additional inverters can be placed between the first inverter (Figure 5-6(b)) and the clamping NMOSFET. During normal operation, the node between the first and second inverter remains at 0 potential and is difficult to charge beyond the trip point of the second inverter. The structure provides excellent noise immunity as a result. Figure 5-7 shows the schematics of a three-stage transient power ESD clamp [22].

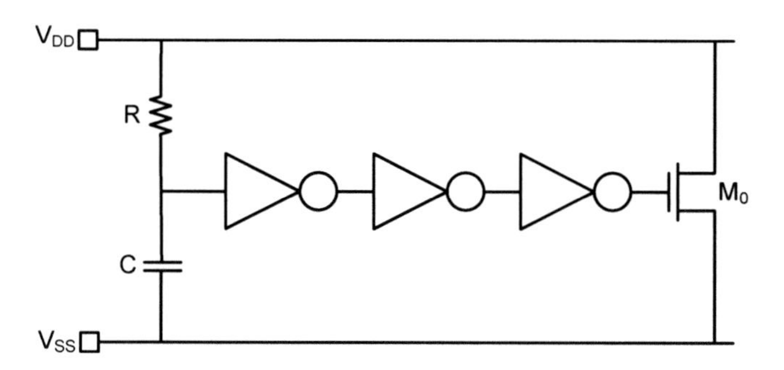

Figure 5-7. Concept of three-stage transient power ESD clamp. (Adapted from [22].)

The system power up under nominal condition has a rise time of the order of milliseconds, while the ESD even will power up in nanoseconds. Therefore, if RC time constant is chosen much larger than the ESD event and much smaller than the nominal power up event, the circuit will be able to distinguish between the two. For example, if the RC = 100 ns, the ESD event will cause all the voltage to appear across R and gate of M_0 will see higher voltage and circuit will trigger. On the other hand, an event with millisecond rise time will have ample time to charge the capacitor C, and M_0 will not trigger. It must be mentioned that the trigger circuit can also be implemented using C-R as shown in Figure 5-8.

Even though, a three inverter transient clamp provides excellent noise immunity it suffers from another problem. In practice, it was found that transient power ESD clamps that include chain of inverters may oscillate during an HBM ESD event and/or normal power-up conditions [19, 32]. This issue has been observed as a high frequency oscillation on the power rails. In order to understand the nature of these oscillations, consider the transient clamp shown in Figure 5-8.

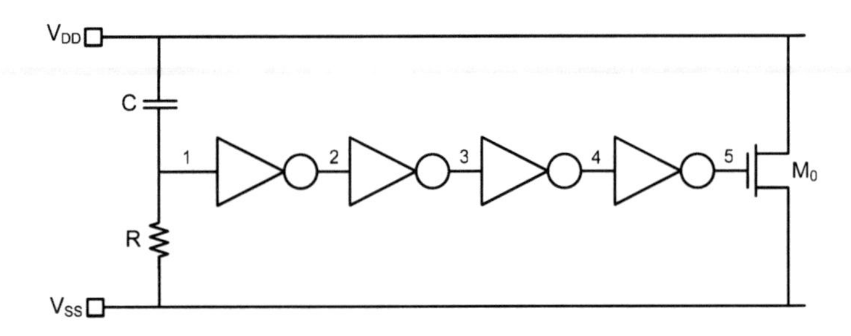

Figure 5-8. Schematic of the second implementation of the transient power ESD clamp.

Under ESD conditions the voltage of the V_{DD} line increases suddenly. Due to the capacitor C_C this jump is transferred to the node 1 as well. Going through four inverters, the gate voltage of the main transistor increases and turns on the clamp transistor. As the ESD energy starts to decay, voltage of the V_{DD} line starts to decrease. The voltage of the node 1 decreased proportionately since voltage across capacitance can not change instantaneously. As the voltage of node 1 reaches the triggering voltage of the first inverter, the inverter chain turns off the main clamp transistor. As a result, voltage of the V_{DD} line increases again, which increases voltage of the node 1 again, and the cycle repeats which results in oscillation. The resulting oscillations on the V_{DD} line are shown in Figure 5-9. It should be noted that these oscillations can be ignored as long as their magnitude is not more than oxide breakdown voltage and they have a limited duration.

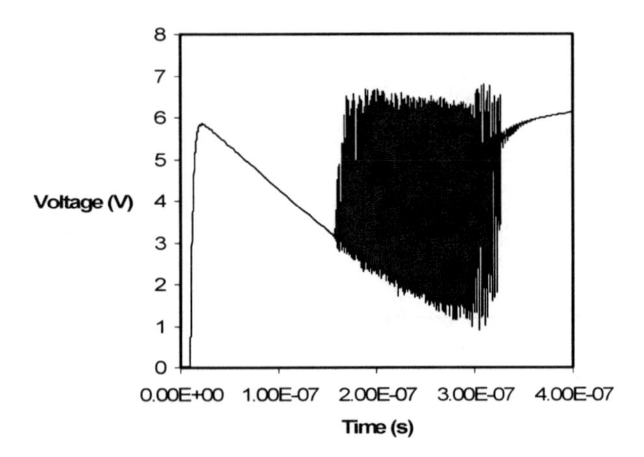

Figure 5-9. Voltage of the V_{DD} line of the transient clamp during 2 kV HBM stress. (Adapted from [32].)

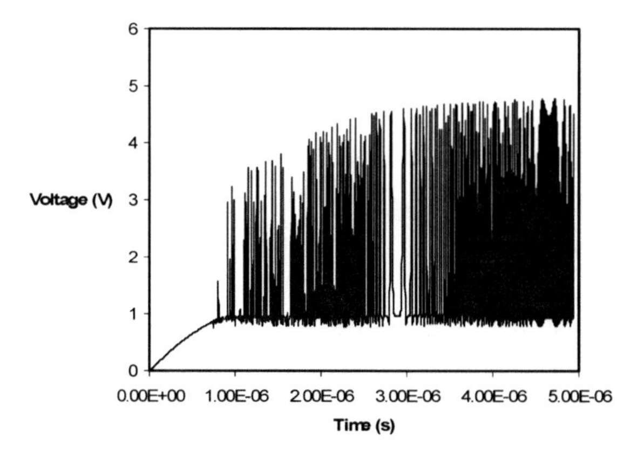

Figure 5-10. Voltage of the V_{DD} line of the transient clamp during a 3 μs power-up event. (Adapted from [32].)

Similar phenomenon can happen during power-up as well. Figure 5-10 shows the oscillation in an inverter-based clamp for a 3 μs rise time power-up condition. The frequency and existence of this oscillation is a function of the rise time of power-up as well as the capacitive load of the power line. Unlike the oscillations under ESD conditions, the oscillations under normal power-up can cause serious issues in normal operating conditions of the main circuit and they should be avoided at all costs.

Stability considerations in clamps are important and must be analyzed before using them [32]. Referring back to oscillation theory, the condition of oscillation is based on the open loop gain of the clamp. The loop is unstable when the magnitude of the open loop gain is 1 and the phase of the open loop gain is 180°. In transient clamps the loop is closed through the power supply rail. Due to the logic of the transient clamps an odd number of inversions (including the R_C-C_C network) exist in the loop. Hence, the condition of 180° phase is satisfied and in order to stabilize the loop the magnitude of the open loop gain must be kept below 1. In order to simulate the open loop gain of the conventional clamp, the feedback loop is opened at the input of the inverter and the impedance seen from each side is added to the other side. Figure 5-11 shows the setup for simulating the open loop gain of the inverter-based clamp. In this figure, C_1 is the input capacitance of the first inverter and the loop gain is defined as V_{out}/V_{in}. By running an AC simulation in Cadence the magnitude and phase of the loop gain is evaluated. Figure 5-12 shows the magnitude and phase of the open loop gain.

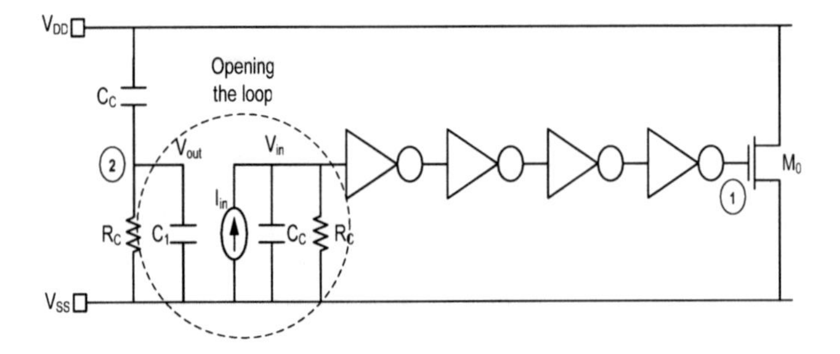

Figure 5-11. Setup to simulate open loop gain of the transient clamp.

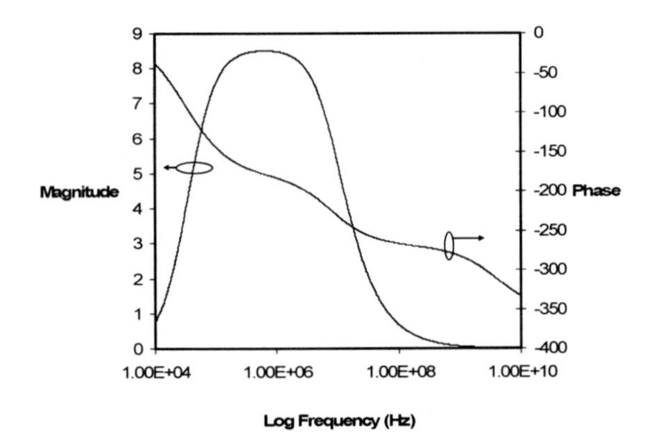

Figure 5-12. Magnitude and phase of the open loop gain of the transient clamp. (Adapted from [32].)

It can be seen that for the conventional clamp a possibility of oscillation exists where the magnitude and phase of the loop gain are 8.49 and –180° respectively. Magnitude of the open loop gain can be used as a measure to ensure immunity to oscillation for different transient clamps [32].

3.3 SRAM-Based ESD Power Clamp

A major limitation of traditional MOSFET based supply protection circuits is the large RC network needed to trigger the clamp and keep it in a conductive state for the entire duration of the ESD pulse. To reduce the RC time constant, the transient power ESD clamp with feedback enhanced triggering was proposed [23]. This clamp enables very fast power supply ramp times and reduces susceptibility to power bus noise. The proposed protection circuit has the embedded SRAM latch as shown in Figure 5-13, which is used as a delay element. The latch stays in "on" mode longer than the RC time constant and provides the sufficient time to dissipate the ESD pulse energy. It was shown that ~25 ns [24] or ~40 ns [28] of RC time is enough to provide the reliable operation of ESD clamp at 2 kV HBM ESD stress. This is extremely beneficial since the RC network can be implemented with greatly reduced physical area.

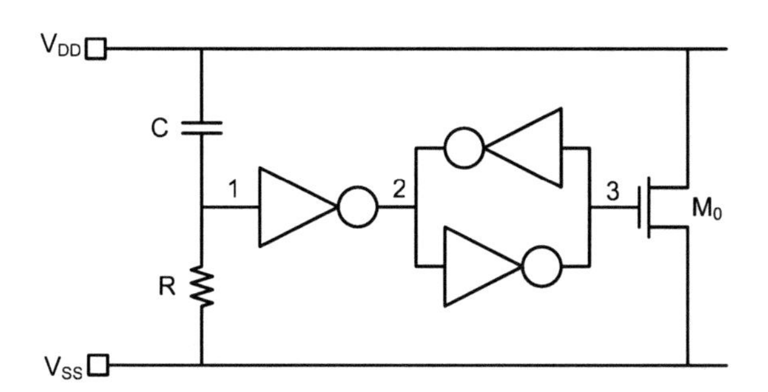

Figure 5-13. SRAM based power ESD clamp with feedback triggering mechanism. (Adapted from [23].)

The isothermal SPICE simulations and TLP measurements of SRAM based power clamp shown that there is a good agreement between simulated and measured *I-V* curves up to 3 kV in 90 nm CMOS technology [23]. However, the recent investigation of this clamp shown that if the impact ionization and thermal effects are not included in the analysis (as would be the case with a SPICE simulation), current flow through the protection device may be greatly underestimated due to the neglecting of parasitic bipolar action in ESD N-MOSFET [24]. As a result, the size of ESD device, required for the given ESD robustness, can be significantly overestimated. Hence, to provide the sufficiently accurate analysis of SRAM based ESD clamps, physical mixed-mode (device-circuit) simulations should be used.

It has been shown that the SRAM delay element keeps the clamp "on" for 700 ns. Since the duration of an ESD event can be as high as 1 μs [28], in order to prevent ESD failure during multiple stress conditions [31], the delay of this clamp should be higher than 1 μs. The delay of this clamp can be further increased by adding an internal capacitive feedback to the triggering circuit. Figure 5-14 shows the schematic of ESD clamp used for mixed-mode (device-circuit) HBM simulations. In this circuit, each transistor is modeled as a physical device structure implemented in 0.18 μm CMOS technology.

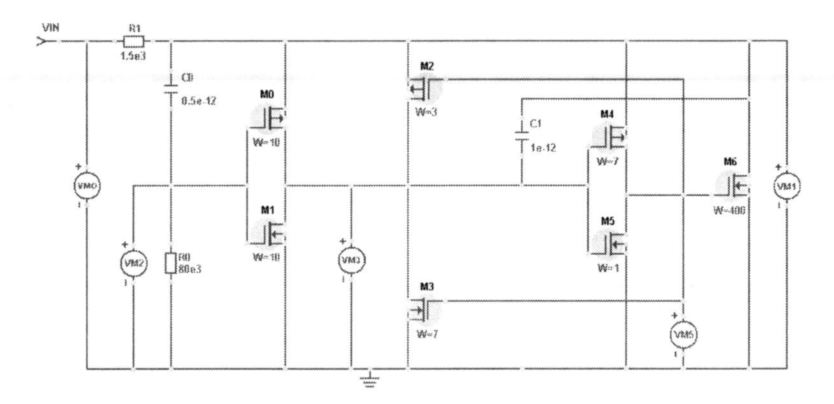

Figure 5-14. Schematics of SRAM based ESD power clamp used for mixed-mode simulations.

The simulation results of SRAM based ESD clamp under 2 kV HBM ESD stress are presented in Figure 5-15. From this figure we can conclude that during the ESD pulse, the potential of RC node (VM2) is raised to high voltage for about 40 ns, which is the time constant of RC network. During this time, the inverter output (VM3) is forced low. It causes the SRAM latch to turn on and keep the gate of ESD MOSFET (M6) at voltage higher than its threshold voltage (V_{th}). Note, that the SRAM latch was designed non-symmetrical (transistors M3 and M4 had bigger size than transistors M2 and M5) to maintain longer the high voltage on the gate electrode of ESD MOSFET (M6). As a result, the ESD clamp is triggered and the output voltage (VM1) is going down after the peak at ~6.5 V.

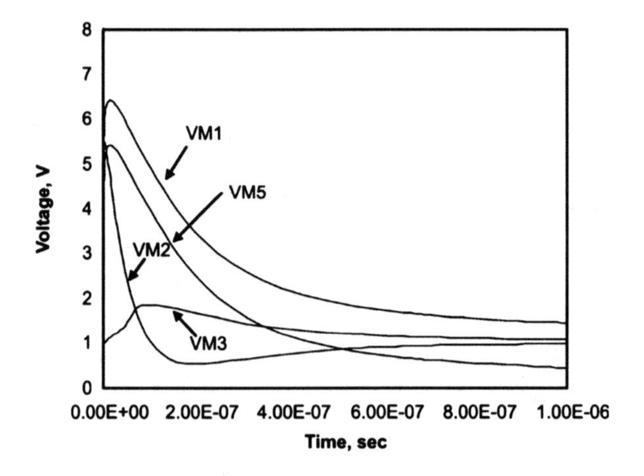

Figure 5-15. Internal behavior of SRAM based ESD clamp at 2 kV ESD stress.

Since the delay element keeps the clamp in "on" mode for over 1 μs, the voltage drop of all nodes decreases after the peak until the ESD energy is completely dissipated, as it shown in Figure 5-15. This peak voltage is low enough to protect the core circuit implemented in 0.18 μm CMOS technology from the 2 kV HBM ESD stress. Note, that the gate oxide breakdown voltage in 0.18 μm CMOS technology at 100 ns voltage pulse is approximately 10 V [25].

Here, it's necessary to remember that the decay time of HBM ESD waveform is 130–170 ns (JESD22-A114-B standard) and the total duration of HBM ESD event is approximately 700–800 ns. The decay time is defined as the time for the waveform to drop from 100% to 36.8% of peak current. It means that only 60–70% of HBM ESD energy is dissipated during the decay time. Hence, to avoid the reliability problems in thin gate oxide core transistors during the HBM ESD event in sub-100 nm CMOS technologies, the transient ESD power clamp must stay in "on" mode during the whole ESD event to completely dissipate ESD energy.

3.4 Thyristor-Based ESD Power Clamp

The analysis of operation of transient ESD clamps have shown that the incorporation of a regenerative feedback network, such as the SRAM latch, can be used to significantly reduce the value of the RC time constant needed to trigger the clamp. It directly translates into reduced circuit area and desired electrical performance during the ESD and system operation. To farther reduce the area occupied by the ESD clamp and to increase its efficiency, the thyristor based ESD clamp was developed [32]. It includes the CMOS thyristor circuit as a delay element [27].

Previously, it was shown that CMOS thyristor circuit can provide the delay in tens of nanoseconds range. It has a low sensitivity to environmental conditions such as supply voltage and temperature and it has low static power consumption [26]. The general concept of proposed transient ESD clamp is depicted in Figure 5-16 [27]. The inverter in this circuit is used for the thyristor triggering, which stays in "on" mode longer than the time constant of conventional RC network and SRAM latch. The schematic of ESD clamp used for mixed-mode (device-circuit) 2 kV HBM ESD TCAD simulations is shown in Figure 5-17. In this circuit, each transistor is modeled as a physical device structure implemented in 0.18 μm CMOS technology.

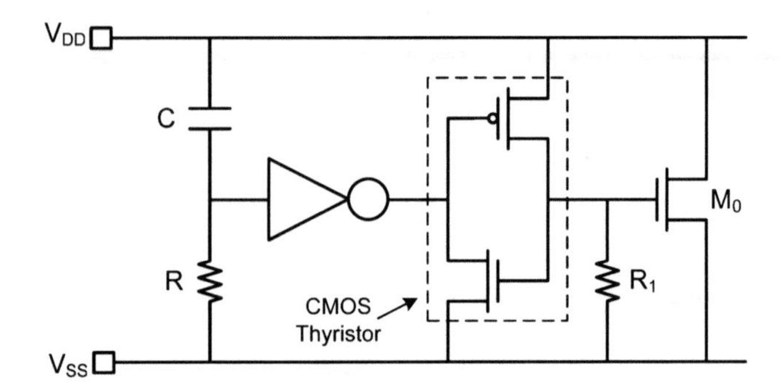

Figure 5-16. Concept of transient ESD clamp with CMOS thyristor delay element.

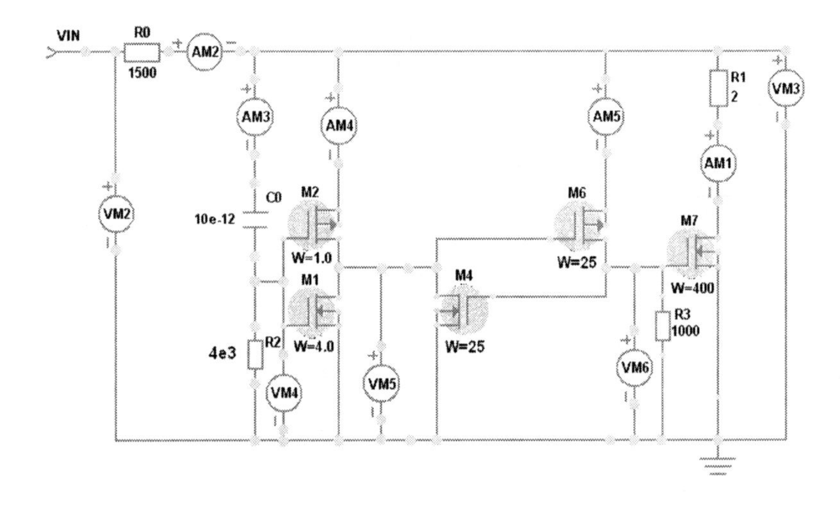

Figure 5-17. Schematics of thyristor based ESD power clamp used for mixed-mode simulations.

The simulation results demonstrate that the thyristor based ESD clamp provides the reliable protection of internal core circuit with the same *RC* time constant (~40 ns) and ESD MOSFET size (400 μm) as a conventional SRAM based ESD clamp without additional feedback capacitor. At 2 kV ESD stress, the peak output voltage was approximately 6 V as shown in Figure 5-18, which is less than the gate oxide breakdown voltage (~8 V). The thyristor based delay circuit stays in the "on" mode for over 1 μs. This time is enough to dissipate whole ESD pulse energy at 2 kV HBM stress.

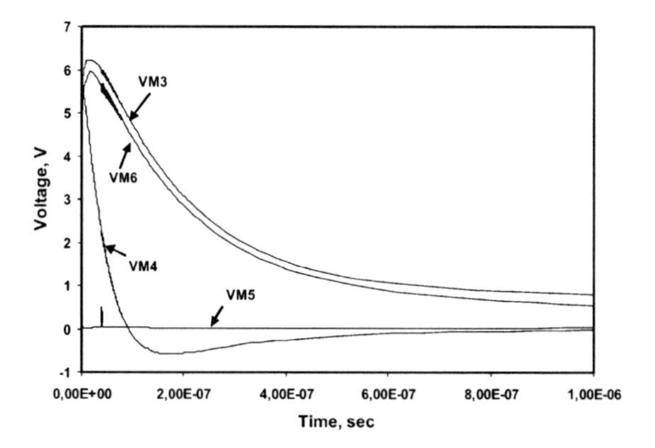

Figure 5-18. Internal behavior of thyristor based ESD power clamp at 2 kV ESD stress (mixed-mode simulation).

3.4.1 Thyristor-Based ESD Power Clamp: Circuit Level Simulations

In advanced CMOS technologies, ESD power clamps the delay element should keep the clamp in "ON" mode during the whole ESD event (between 500 ns and 1 μs) [28]. In order to evaluate the effectiveness of non-snapback clamp under whole ESD event, it was simulated in circuit simulation environment using 0.18 μm CMOS technology with t_{ox} = 40 Å (gate oxide thickness). The width of the main discharging transistor (M7) shown in Figure 5-17 was set to 400 μm, which was realized in a 40 finger configuration. The transistors (M4 and M6) of the thyristor delay element were 10 μm width each. The resistor R3 is used to turn off the main clamp transistor M7 under normal operating conditions. Figure 5-19 shows the voltage

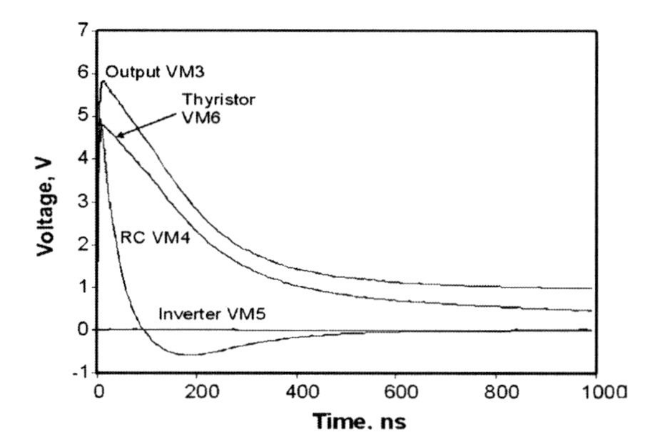

Figure 5-19. ESD response of thyristor based ESD power clamp at a 2 kV HBM event (circuit simulation).

of different nodes of thyristor based power clamp at a 2 kV HBM ESD stress simulated in a circuit simulation environment. Referring to Figure 5-18 it can be seen that the voltage of VM5 is kept at '0' during the ESD event keeping the clamp "on" for over 1 μs. Width of M7 determines the peak of the supply rail, which is 5.8 V in our design. Since the gate oxide breakdown voltage in 0.18 μm CMOS technology under 100 ns voltage pulse is around 8 V [29], hence this clamp can effectively protect the core circuit against a 2 kV ESD event. Note that the results of circuit-level simulations performed in Cadence are similar to the mixed-mode TCAD simulation results shown in Figure 5-18 [32].

3.4.2 Immunity to False Triggering

In addition to the ESD response, the thyristor based power ESD clamp should be tested under normal operating conditions as well. The immunity to false triggering is evaluated by applying a ramp from 0 to V_{DD} with different rise times. In regular applications, power-up is a very slow event where the rise time is in the millisecond range. However, in some applications such as hot plug operations, the rise time can be around 1 μs that may cause false triggering. In order to investigate the worst case scenario, the non-snapback ESD power clamp was simulated in Cadence for a 1 μs of power-up. Figure 5-20 shows the voltage drop of different nodes for this experiment [32].

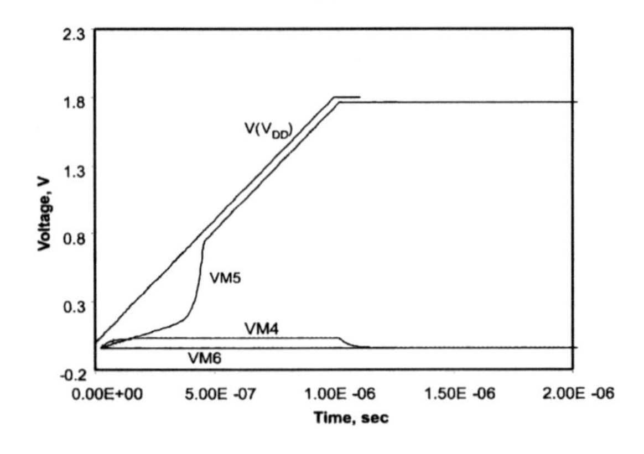

Figure 5-20. Transient behavior of thyristor based ESD power clamp at 1 μs power-up event.

It can be seen that the voltage of RC node VM4 increases up to 70 mV only, while the voltage of node VM6 (see Figure 5-17) is almost zero all the time. Hence, even a very fast power-up signal (1 μs) will not trigger the clamp and the design of thyristor based power ESD clamp is immune to false triggering.

3.4.3 Immunity to Power Supply Noise

As chip power densities and power consumption is on the rise, the power supply noise is becoming an important issue. In the context of ESD power supply clamps, the clamp should be robust against such noises. This issue becomes an important factor in circuits with high switching rates. Immunity to power supply noise can be evaluated by adding a pseudo-random pulse of a given magnitude superimposed on to the supply voltage. In this sub-section we consider a noise of 600 $mV_{p\text{-}p}$ at 500 Mbps added to the supply line. Figure 5-21 shows the gate voltage of ESD transistor M7 and its current under these conditions [32].

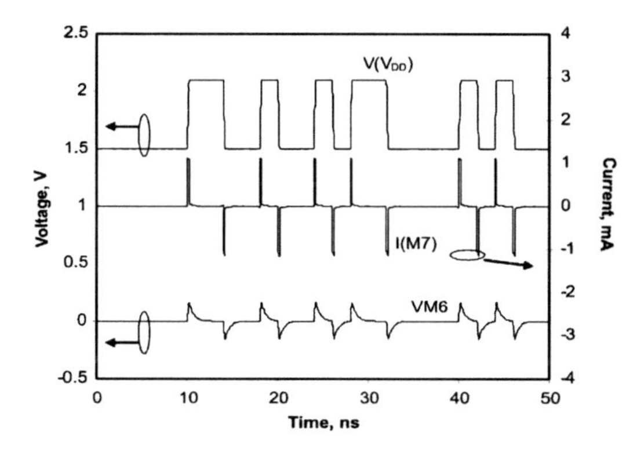

Figure 5-21. Power supply noise immunity: supply voltage V(VDD), gate voltage of transistor.

It can be seen that the peak voltage of M7 is around 0.25 V. It is significantly lower than the threshold voltage of ESD transistor M7 (V_{th} = 0.5 V). Hence, the thyristor based ESD power clamp does not turn on and is immune to the power supply noise.

3.4.4 Immunity to Oscillation

In order to test the stability of the thyristor based clamp, similar to the method explained in Section 3.2, the loop is opened from node 1. Then the impedance seen from each side is added to the other side. The magnitude of the loop gain V_{out}/V_{in} should be less than 1 to ensure stability. Figure 5-22 shows the magnitude and phase of the loop gain. It can be seen that in this clamp the magnitude of the loop gain is always less than 1. Therefore, this clamp is immune to oscillation.

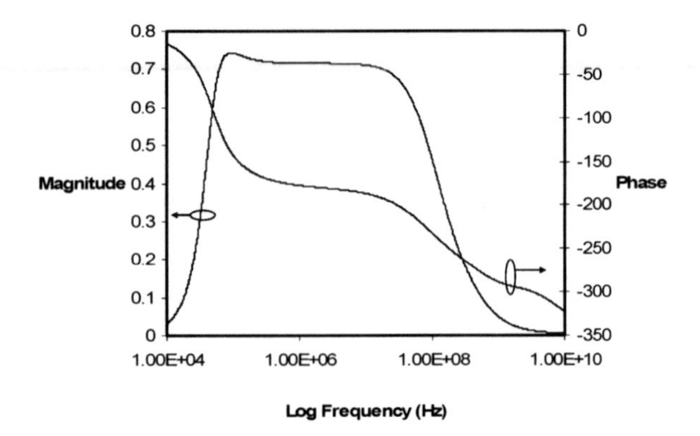

Figure 5-22. Magnitude and phase of the loop gain of the thyristor-based clamp.

3.4.5 TLP and HBM Measurements

TLP measurements of thyristor based ESD power clamp fabricated in 0.18 μm CMOS technology were performed using Pulsar 900 TLP system [30]. Figure 5-23 shows the TLP measurement results using 100 ns wide pulses with 10 ns rise time.

It can be seen that the leakage current at V_{DD} = 1.8 V is 7 nA and the second breakdown current is 1.83 A. Additionally, HBM tests have been done on the clamp as well. These measurements were performed using the IMCS-700 HBM/MM ESD tester. The positive and negative HBM stresses with 500 V step sizes were applied. This clamp passed both +3 kV and –3 kV stresses.

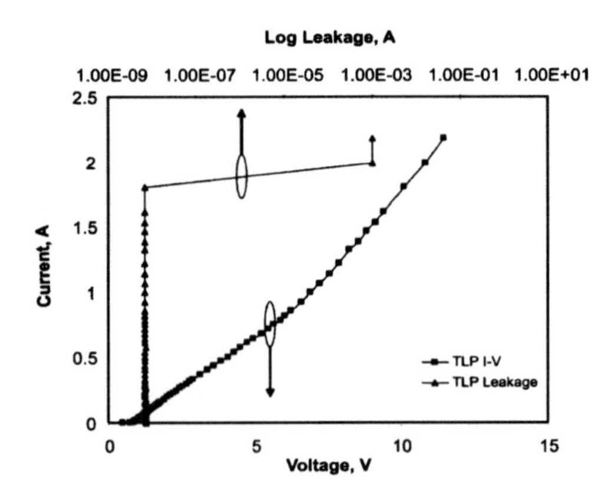

Figure 5-23. TLP measurement results of thyristor based ESD power clamp.

3.5 Flip-Flop-Based Transient Power Supply Clamp

As mentioned earlier, the triggering circuit is divided into two parts: the rise time detector and the delay element. This delay element should be designed to keep the main transistor (M_0) in "on" state for the entire duration of the ESD event. In this sub-section we consider a delay element based on a rising edge triggered D-type flip-flop. Figure 5-24 shows the schematic of flip-flop based power supply ESD clamp [33]. In order to turn off the clamp in normal operating conditions, the gate of M_2 should be connected to 'clk'. As a result, at normal conditions where V_{DD} is connected to power supply and C_C is fully charged, the transistor M_2 is turned on and transistor M_0 is turned off, respectively. In order to ensure a proper operation, M_1 should be designed to be larger than M_2 and M_3 to be able to pull up the input of the inverter under normal operating conditions.

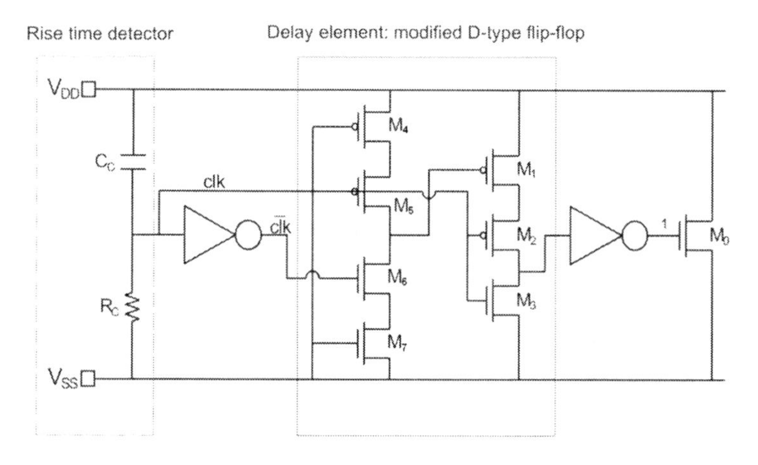

Figure 5-24. Schematic of the flip-flop based power supply ESD clamp [33].

One of the key features of the non-snapback ESD protection scheme is the possibility to simulate the clamp using circuit level simulation tools such as Cadence. This feature is based on the fact that the transistors of the clamp are not operating in their breakdown region. In order to simulate this clamp with Cadence, the 2 kV HBM ESD pulse was applied to the V_{DD} line with grounded V_{SS}. The circuit, shown in Figure 5-24, was simulated using 0.18 μm CMOS technology with t_{ox} = 40 Å (the gate oxide thickness of MOS transistors). The transistor M_0 was designed with the minimum length and 400 μm width. Figure 5-25 depicts the voltage of different nodes under ESD stress.

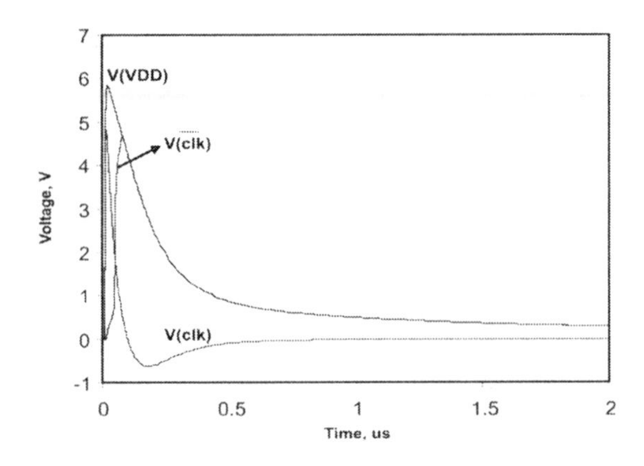

Figure 5-25. ESD response of flip-flop based ESD power clamp at a 2 kV HBM event (circuit simulation).

Based on the voltage of the V_{DD} node, it can be seen that the clamp is "on" for over 2 μs and until the ESD event decays completely. The peak voltage of the supply line during the 2 kV stress is 5.8 V. Therefore, flip-flop based ESD clamp can effectively protect the core circuit from a 2 kV ESD event.

3.5.1 Immunity to False Triggering

In normal operational conditions, the system power supply will ramp at a predefined rate, which is usually in the range of several milliseconds to several tens of milliseconds. This is several orders of magnitude slower than the power supply rise times seen during ESD events. In response to the slowly increasing voltage rate on the rail V_{DD}, the RC node in Figure 5-23 remains at a potential near ground, since the R_c resistor can effectively remove the charge deposited by C_c capacitor. However, in some applications such as hot plug operations the rise time can be as fast as 1 μs that may cause the false triggering of ESD power clamp. Hence, to test the worst case scenario, the flip-flop based power clamp should be simulated at a 1 μs power-up. In order to avoid false triggering the gate voltage of M_0 (node 1) should be less than its threshold voltage ($V_{th} \approx 0.45$ V). Figure 5-26 shows the voltage of different nodes for a 1 μs power-up. It can be seen that the voltage of node 1 (the gate terminal of transistor M_0) rises to 0.12 V and goes back to 0 immediately. Hence, even with a very fast power-up event, this clamp doesn't trigger which ensures the immunity to false triggering.

As mentioned earlier, the concept of using a flip-flop to latch the gate of M_0 to '1' under ESD conditions brings the concern of turning off the clamp

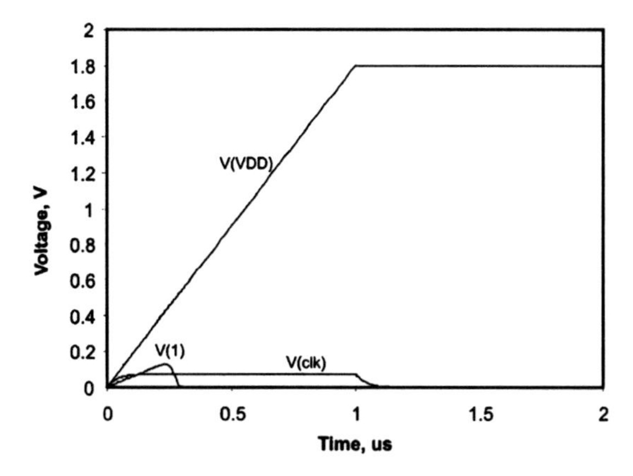

Figure 5-26. Transient behavior of flip-flop based ESD power clamp at 1 μs power-up event.

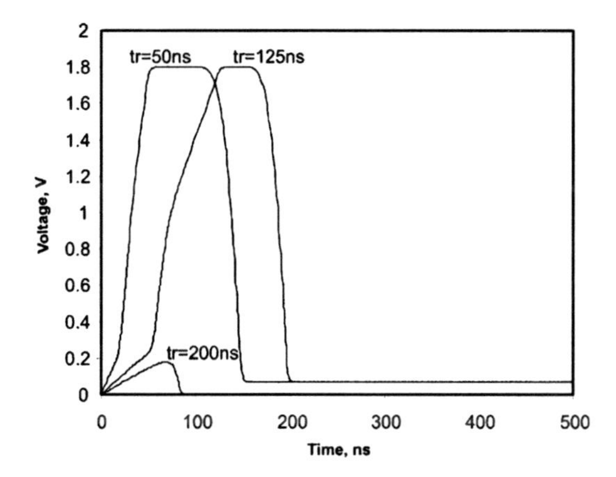

Figure 5-27. The voltage of node 1 (M_0) for tr = 50, 125 and 200 ns.

after triggering. Therefore, another set of simulations is necessary to make sure the clamp turns off after the possible false triggering. To investigate this, the rise time of the power-up event is further reduced to 200, 125 and 50 ns. Figure 5-27 shows the voltage of node 1 during these power-up events.

It can be seen that for the higher rise time of 200 ns the clamp doesn't trigger. On the other hand, for smaller rise time of 125 and 50 ns the clamp turns on at the power-up event but turns off after less than 50 ns. This simulation ensures that the turn off mechanism of the clamp under normal operating conditions is effective.

3.5.2 Immunity to Power Supply Noise

As it was mentioned earlier, the power supply noise becomes an important factor in circuits with high switching rates. To investigate the impact of power supply noise on the clamp, a pseudo-random pulse is added to the supply voltage. In order to simulate the clamp in worst case conditions the added noise has a rate of 500 Mbps and the amplitude of 600 $mV_{p\text{-}p}$ [23]. This bit sequence along with the voltage of node 1 (the gate voltage of transistor M_0) is shown in Figure 5-28. It can be seen that the peak voltage of M_0 is around 0.15 V. It is a significantly lower than the threshold voltage of ESD transistor M_0 (V_{th} = 0.5 V). Hence, the flip-flop based ESD power clamp is immune to the power supply noise.

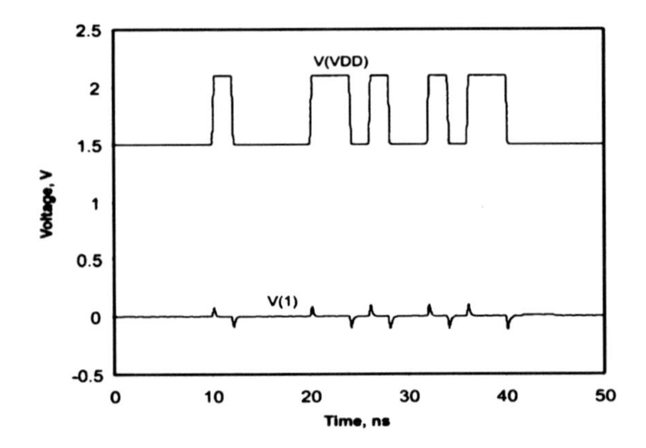

Figure 5-28. Power supply noise immunity: supply voltage $V(V_{DD})$ and gate voltage of transistor M_0 (V(1)).

3.5.3 Immunity to Oscillation

In order to test the stability of the flip-flop based clamp, similar to the method used in Section 3.2, the clamp is opened at node clk and the impedance seen from each side is added to the other side. Figure 5-29 shows the magnitude and phase of the open loop gain for this clamp. It can be seen that the magnitude of the loop gain of the proposed clamp is always less than 1 and hence, immune to oscillation.

3.5.4 TLP and HBM Measurements

The flip-flop based ESD power clamp has been fabricated in 0.18 μm CMOS TSMC technology. TLP measurements have been done using the Pulsar 900

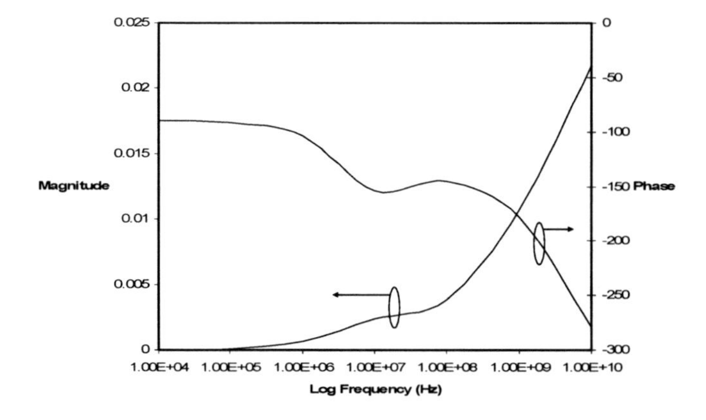

Figure 5-29. Magnitude and phase of the loop gain of the flip-flop based clamp.

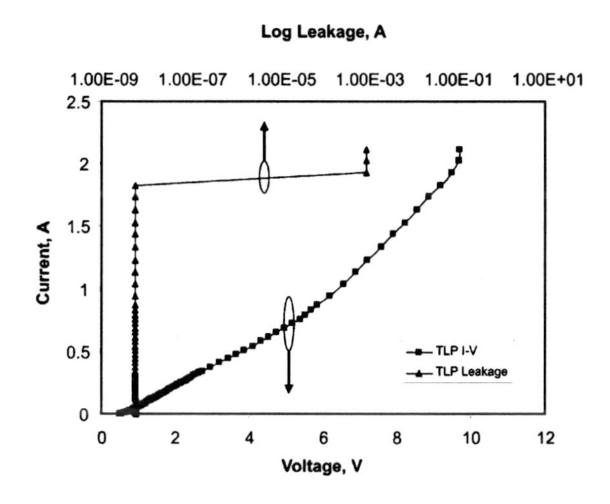

Figure 5-30. TLP measurement results of flip-flop based ESD power clamp.

TLP system from SQP Products [30]. Figure 5-30 shows the TLP measurement results for this clamp using 100 ns wide pulses.

It can be seen that the leakage current at V_{DD} = 1.8 V is 6 nA and the second breakdown current is approximately 1.83 A. Furthermore, HBM test has been done on the clamp as well. These measurements were done using IMCS-700 HBM/MM ESD tester. We applied both positive and negative HBM stresses with 500 V step sizes. This clamp passed both +3 and –3 kV stresses. But when we increased the stress to 3.5 kV it passes +3.5 kV stress while fails –3.5 kV stress.

4. SUMMARY

The ESD protection methods require the placement of adequate on-chip protection devices in the I/O frame and in the power supply pins to absorb the ESD energy. The on-chip protection scheme should have an explicitly robust path for the ESD currents which flow between any pair of pins. In general, pad protection networks shunt I/O pins to the ground bus under stress events. For each input pin, a dedicated protection network that is completely passive under usual operating conditions has to be added. For each output pin, the ESD protection level is determined by the intrinsic robustness of the output buffer transistors plus that of the dedicated protect-tion device.

In order to develop the effective whole-chip ESD protection, the ESD protection circuits in input and output pads are not enough. The effective ESD clamp circuits between the power rails are necessary for protecting the core circuits against the ESD damage. ESD power clamps are typically grouped into two categories: static and transient. The static power clamps include ESD circuits based on diode strings, MOS transistors and SCR devices. The transient power clamps have the RC network for triggering of ESD MOSFETs or SCR devices. The three-stage inverters, SRAM and thyristor based transient ESD power clamps were developed and practically used in industry. In this chapter, the static and transient power clamps were analyzed. The robustness of transient power clamps at normal operating conditions was also discussed.

REFERENCES

[1] S. H. Voldman, G. Gerosa, V. P. Gross, N. Dickson, S. Furkay, and J. Slinkman, "Analysis of snubber-clamped diode-string mixed voltage interface ESD protection network for advanced microprocessors," *EOS/ESD Symposium*, pp. 43–61, 1995.

[2] T. J. Maloney and S. Dabral, "Novel clamp circuits for IC power supply protection," *EOS/ESD Symposium*, pp. 1–12, 1995.

[3] S. Dabral, R. Aslett, and T. Maloney, "Designing on-chip power supply coupling diodes for ESD protection and noise immunity," *EOS/ESD Symposium*, pp. 239–249, 1993.

[4] J. C. Bernier, G. D. Croft, and W. R. Young, "A process independent ESD design methodology," *Proc. of the Int. Symp. on Circuits and Systems VLSI*, pp. 218–221, 1999.

[5] M. -D. Ker and W. -Y. Lo, "Design on the low-leakage diode string for using in the power-rail ESD clamp circuits in a 0.35-μm silicide CMOS process," *IEEE J. of Solid-State Cir.*, vol. 35, No. 4, pp. 601–11, 2000.

[6] T. J. Maloney, K. K. Parat, N. K. Clark, and A. Darwish, "Protection of high voltage power and programming pins," *IEEE Trans. on Components, Packaging & Manufacturing Technology, Part C (Manufacturing)*, vol. 21, No. 4, pp. 250–256, 1998.

[7] W. Wondrak, "Physical limits and lifetime limitations of semiconductor devices at high temperature," *Microelectronics Reliability*, vol. 39, No. 6–7, pp. 1113–1120, 1999.
[8] S. Dabdal, R. Aslett, and T. Maloney, "Designing on chip power supply coupling diodes for ESD protection and noise immunity," *Journal of Electrostatics*, vol. 33, No. 3, pp. 357–370, 1994.
[9] S. Dabral and T. J. Maloney, *Basic ESD and I/O Design*, Wiley, New York, 1998.
[10] T. -S. Yeoh, "ESD effects on power supply clamps," *Int. Symp. on the Physical and Failure Analysis of Integrated Circuits (IPFA)*, pp. 121–124, 1997.
[11] C.Richer, N. Maene, G. Mabboux, and R. Bellens, "Study of the ESD behavior of different clamp configurations in a 0.35 μm CMOS technology," *EOS/ESD Symposium*, pp. 240–245, 1997.
[12] M. -D. Ker and T. -Y. Chen, "Design on the turn-on efficient power-rail ESD clamp circuit with stacked polysilicon diodes," *Proc. of the Int. Symp. on Circuits and Systems (ISCAS)*, pp. 758–761, 2001.
[13] M. -D. Ker; C. -Y. Wu, T. Cheng, M. J. -N. Wu, T. -L. Yu, and A.C. Wang, "Whole-chip ESD protection for CMOS VLSI/ULSI with multiple power pins," *Proc. of the Int. Integrated Reliability Workshop*, pp. 124–128, 1994.
[14] M. -D. Ker and H. -H. Chang, "How to safely apply the LVTSCR for CMOS whole-chip ESD protection without being accidentally triggered on," *EOS/ESD Symposium*, pp. 72–85, 1998.
[15] M. P. J. Mergens, C. C. Russ, K. G. Verhaege, J. Armer, P. C. Jozwaik, and R. Mohn, "High holding current SCRs (HHI-SCR) for ESD protection and latch-up immune IC operation," *EOS/ESD Symposium*, paper 1A3, 2002.
[16] M. P. J. Mergens, C. C. Russ, K. G. Verhaege, J. Armer, P. C. Jozwiak, R. Mohn, B. Keppens, and C. S. Trinh, "Diode-triggered SCR (DTSCR) for RF-ESD protection of BiCMOS SiGe HBTs and CMOS ultra-thin gate oxides," *Proc. of the Int. Electron Devices Meeting (IEDM)*, pp. 515–518, 2003.
[17] M. Mergens, G. Wybo, B. Van Camp, B. Keppens, F. De Ranter, K. Verhaege, P. Jozwiak, J. Armer, and C. Russ, "ESD protection circuit design for ultra-sensitive IO applications in advanced sub-90 nm CMOS technologies," *Proc. of the Int. Symp. on Circuits and Systems (ISCAS)*, pp. 1194–1197, 2005.
[18] V. A. Vashchenko, A. Concannon, M. ter Beek, and P. Hopper, "High holding voltage cascoded LVTSCR structures for 5.5-V tolerant ESD protection clamps," *IEEE Trans. on Device and Materials Reliability*, vol. 4, No. 2, pp. 273–280, 2004.
[19] B. L. Hunter and B. K. Butka, "Damped transient power clamps for improved ESD protection of CMOS," *Microelectronics Reliability*, vol. 46, No. 1, pp. 77–85, 2006.
[20] R. Merrill and E. Issaq, "ESD design methodology," *Proc. of EOS/ESD Symposium*, pp. 223–237, 1993.
[21] J. C. Bernier, G. D. Croft, and W. R. Young, "A process independent ESD design methodology," *Proc. of the Int. Symp. on Circuits and Systems (ISCAS)*, pp. 218–221, 1999.
[22] R. Merrill and E. Issaq, "ESD design methodology," *Proc. of EOS/ESD Symposium*, pp. 223–237, 1993.
[23] J. C. Smith and G. Boselli, "A MOSFET power supply clamp with feedback enhanced triggering for ESD protection in advanced CMOS technologies," *Microelectronics Reliability*, vol. 45, No. 2, pp. 201–210, 2005.
[24] V. Axelrad and A. Shibkov, "Mixed-mode circuit-device simulation of ESD protection circuits with feedback triggering," *Int. Conf. on Solid-State Device and Materials (SSDM)*, paper P1-2, 2004.

[25] A. Salman, R. Gauthier, E. Wu, P. Riess, C. Putnam, M. Muhammad, Min Woo, and D. Ioannou, "Electrostatic discharge induced oxide breakdown characterization in a 0.1 μm CMOS technology," *Proc. of the Int. Reliability Physics Symposium (IRPS)*, pp. 170–174, 2002.

[26] G. Kim, M.-K. Kim, B.-S. Chang, and W. Kim, "A low-voltage, low-power CMOS delay element," *IEEE J. on Solid-State Cir.*, vol. 31, No. 7, pp. 966–971, 1996.

[27] H. Sarbishaei, O. Semenov, and M. Sachdev, "A transient power supply ESD clamp with CMOS thyristor delay element," *Proc. EOS/ESD Symposium*, pp. 395–402, 2007.

[28] M. Stockinger, J. W. Miller, M. G. Khazhinsky, C. A. Torres, J. C. Weldon, B. D. Preble, M. J. Bayer, M. Akers, and V. G. Kamat, "Boosted and distributed rail clamp networks for ESD protection in advanced CMOS technologies," *Proc. of EOS/ESD Symposium*, pp. 17–26, 2003.

[29] H. Gossner, "ESD protection for the deep sub micron regime-a challenge for design methodology,", *Int. Conf. on VLSI Design*, pp. 809–818, 2004.

[30] http://www.sqpproducts.com

[31] R. A. Ashton, B. E. Weir, G. Weiss, and T. Meuse, "Voltages before and after HBM stress and their effect on dynamically triggered power supply clamps," *Proc. EOS/ESD Symposium*, pp. 153–159, 2004.

[32] H. Sarbishaei, O. Semenov, and M. Sachdev, "A transient power supply ESD clamp with CMOS thyristor delay element," *Proc. EOS/ESD Symposium*, pp. 395–402, 2007

[33] H. Sarbishaei, "Electrostatic Discharge Circuit for High-Speed Mixed-Signal Circuits," Ph.D. Thesis, University of Waterloo, Canada, 2007.

Chapter 6

ESD PROTECTION CIRCUITS FOR HIGH-SPEED I/OS

1. INTRODUCTION

In previous chapters, we discussed design of ESD protection circuits with respect to device and circuit parameters. Similarly, most of the ESD related publications are focused on these aspects as well. However, ESD circuit parameters do influence the design parameters of circuit to be protected. Therefore, the interaction between ESD protection circuit and the main circuit is another challenge that should be studied carefully. This issue is becoming very important as circuits are moving towards higher frequencies and higher data rates since ESD parasitics may affect the performance of these circuits.

An ESD protection circuit adds extra parasitic capacitance to the main circuit. This capacitance is mainly reverse biased *pn* junction capacitance, which is highly non-linear. As a result, an ESD protection circuit can degrade both frequency response and linearity performance of the main circuit. The former, which is due to mere presence of the parasitic capacitance, has been well understood. However, the latter, which is due to non-linear behavior of the junction capacitance, hasn't been discussed in detail. In this chapter, the main focus is on the impact of ESD protection on high performance circuits. Hence, in Section 2 parasitic capacitance associated with different ESD protection circuits is presented. In Sections 3 to 5 three different case studies are discussed and impact of ESD protection circuits on their respective performance is examined. The first

O. Semenov et al., ESD Protection Device and Circuit Design for Advanced CMOS Technologies, 147–172.

two are on the impact of ESD protection circuit on linearity of analog to digital converters (ADCs), while the last one reports the impact of ESD protection on jitter of current mode logic (CML) drivers.

2. PARASITIC CAPACITANCE OF ESD PROTECTION CIRCUITS

ESD protection circuits add parasitic capacitance to the main circuit which limits its performance. Therefore, before discussing the impact of ESD protection on main circuit's performance, it is necessary to study the parasitic capacitance associated with different ESD protection circuits.

Based on discussions in Chapter 3, parasitic capacitance of ESD protection circuits consists of reverse biased *pn* junction capacitance and gate capacitance. For example, in diode and SCR, parasitic capacitance is the reverse biased *pn* junction capacitance only. On the other hand, in GGNMOS, parasitic capacitance has two components: gate capacitance and reverse biased drain-bulk junction capacitance. In the coming two sections *pn* junction capacitance and gate capacitance are discussed.

2.1 Reverse-Biased *pn* Junction Capacitance

Under normal operating conditions, the reverse-biased *pn* junction capacitance is the main parasitic capacitance of ESD protection circuits. Consider the cross section of a *pn* junction diode shown in Figure 6.1(a). In order to simplify the equations, dopings of *n*-type and *p*-type regions are assumed to be constant and equal to N_D and N_A respectively. Consequently, a depletion region is formed at both sides of the junction. Figure 6-1(b) shows the charge density in the depletion region where W_1 and W_2 are widths of the depletion regions in *p*-type and *n*-type regions respectively. As the reverse bias voltage (V_R) increases, width of the depletion region (W) increases as well. In other words, there is a voltage dependent charge associated with the depletion region. Hence, a small signal capacitance is defined for the diode. The value of this capacitance is a function of the width of the depletion region.

First, let's assume V_R is zero. In this case, a voltage ψ_0 is created across the junction that prevents the diffusion of mobile holes and electrons across the junction in equilibrium. This voltage is called built-in potential and is calculated from the following equation [1]:

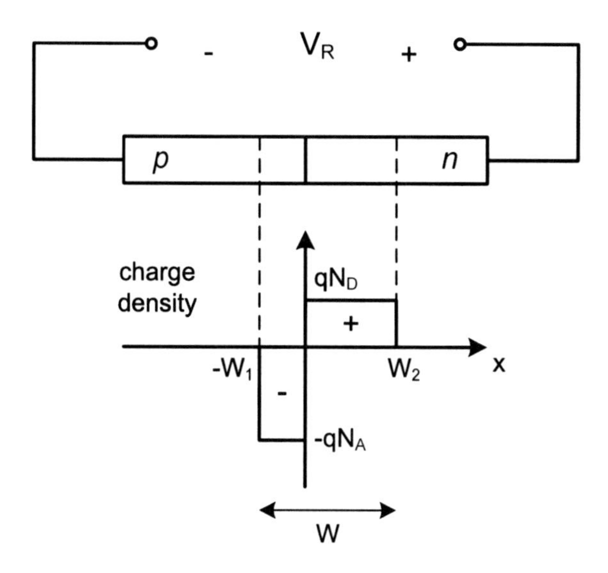

Figure 6-1. Schematic of a diode under reverse bias V_R.

$$\psi_0 = V_T \ln \frac{N_A N_D}{n_i^2} \tag{6-1}$$

In equation (6-1) V_T is the thermal voltage and n_i is the intrinsic carrier concentration. At 300 K, $V_T = 26$ mV and $n_i = 1.5 \times 10^{10}$ cm^{-3}.

Now consider the case where V_R is positive. Referring to polarity of charge in the depletion region, total voltage across the junction becomes $(\psi_0 + V_R)$. As total charge per unit area on either side of the junction should be equal in magnitude, the following equation can be written:

$$W_1 N_A = W_2 N_D \tag{6-2}$$

In order to find width of the depletion region in the diode, Poisson's equation should be solved in both *n* and *p* regions. Solving Poisson's equation in one dimension and in depletion region of the *p*-type semiconductor ($-W_1 < x < 0$) results in the following equation [1]:

$$V = \frac{qN_A}{\varepsilon}\left(\frac{x^2}{2} + W_1 x + \frac{W_1^2}{2}\right) \text{ for } -W_1 < x < 0 \tag{6-3}$$

Equation (6-3) is derived assuming the voltage at $x = -W_1$ to be zero. The voltage at the junction boundary can be calculated from equation (6-3) by setting $x = 0$. This voltage is called V_1 and is shown in equation (6-4):

$$V_1 = \frac{qN_A}{\varepsilon}\frac{W_1^2}{2} \tag{6-4}$$

One dimensional Poisson's equation is solved for the depletion region of the *n*-type semiconductor ($0 < x < W_2$). The voltage difference between $x = 0$ and $x = W_2$, which is called V_2, becomes:

$$V_2 = \frac{qN_D}{\varepsilon}\frac{W_2^2}{2} \tag{6-5}$$

The voltage across the *pn* junction is $V_1 + V_2$ and is calculated by adding equations (6-4) and (6-5):

$$V_1 + V_2 = \frac{q}{2\varepsilon}\left(N_A W_1^2 + N_D W_2^2\right) = \psi_0 + V_R \tag{6-6}$$

Substituting (6-2) into (6-6):

$$\psi_0 + V_R = \frac{qW_1^2 N_A}{2\varepsilon}\left(1 + \frac{N_A}{N_D}\right) \tag{6-7}$$

Equation (6-7) can be rearranged to calculate width of the depletion region in the p-type semiconductor:

$$W_1 = \sqrt{\frac{2\varepsilon(\psi_0 + V_R)}{qN_A\left(1 + \frac{N_A}{N_D}\right)}} \tag{6-8}$$

Finally, by solving W_2 from (6-2) and substituting it in (6-6), width of the depletion region in n-type semiconductor is derived as well:

$$W_2 = \sqrt{\frac{2\varepsilon(\psi_0 + V_R)}{qN_D\left(1 + \frac{N_D}{N_A}\right)}} \tag{6-9}$$

Knowing the width of the depletion region, the capacitance associated with the depletion region is calculated from the following equation:

$$C_j = \frac{dQ}{dV_R} = \frac{dQ}{dW_1}\frac{dW_1}{dV_R} \tag{6-10}$$

The charge in the *p*-type region can be written as:

$$Q = AqN_A W_1 \tag{6-11}$$

By substituting equations (6-8) and (6-11) into (6-10) total capacitance of the *pn* junction is derived:

$$C_j = A\sqrt{\frac{q\varepsilon N_A N_D}{2(N_A + N_D)}}\frac{1}{\sqrt{\psi_0 + V_R}} \tag{6-12}$$

It is common to modify equation (6-12) to include forward biased diodes as well. Hence, instead of V_R, a general term, V_D, is used. V_D is defined as the voltage across anode-cathode of the diode. In forward-biased region, V_D is positive, while in reverse-biased region, V_D in negative. Using this terminology, equation (6-12) is rewritten as follows:

$$C_j = A\sqrt{\frac{q\varepsilon N_A N_D}{2(N_A + N_D)}}\frac{1}{\sqrt{\psi_0 - V_D}} = \frac{C_{j0}}{\sqrt{1 - \frac{V_D}{\psi_0}}} \tag{6-13}$$

where C_{j0} is the junction capacitance when $V_D = 0$.

There are a few points that can be concluded from equation (6-13). As the reverse bias voltage is increased, junction capacitance becomes smaller. Moreover, the junction capacitance has non-linear voltage dependence. As will be discussed in the coming sections, this non-linear dependence is responsible for increasing the distortion of analog circuits such as analog to digital converters.

2.2 Gate Capacitance of MOSFET

Unlike *pn* junction diode, MOSFET has a much more complicated capacitance model, which depends on its operating region. The small signal model shown in Figure 6-2 is widely used to calculate MOSFET's parasitics [1].

In the model shown in Figure 6-2, g_m, g_{mb} and g_{sd} are small signal gate, substrate and source-drain transconductances respectively. The main focus of this section is the capacitance of MOSFETs. Therefore, the details of the small signal model are not discussed in this chapter. However, it's necessary to calculate the value of parasitic capacitances in different operating regions. C_{bd} and C_{bs} are junction capacitances and their equations are similar to equation (6-13). C_{gs}, C_{gd} and C_{gb} are gate capacitances and are discussed in the rest of this section.

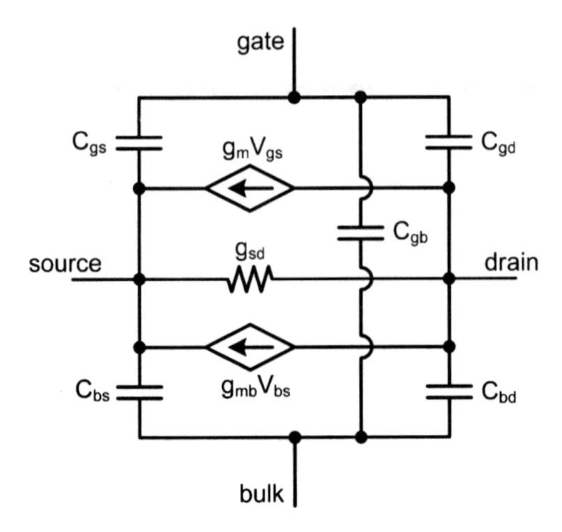

Figure 6-2. Small signal model of a MOS transistor.

In order to simplify the discussion, we consider two operating regions for MOSFETs: weak inversion and strong inversion. Strong inversion is divided into saturation and linear regions.

In weak inversion, the inversion layer charge is negligible in the channel. Hence, the impact of drain (source) voltage on gate capacitance is very small and as a result, C_{gd} (C_{gs}) consists of gate-drain (gate-source) overlap capacitance and fringing capacitance [2]:

$$C_{gs} = C_{gd} = C_{ol} = WL_{ol}C_{ox} \tag{6-14}$$

where L_{ol} is overlap distance and C_{ox} is the oxide capacitance. It should be noted that L_{ol} is usually calculated empirically.

However, C_{gb} is much larger than C_{gs} and C_{gd} and can be considered as the series combination of oxide capacitance and depletion capacitance [1]:

$$C_{gb} = WL\left(\frac{C_{ox}C_{js}}{C_{ox} + C_{js}}\right) \tag{6-15}$$

where C_{js} is the depletion region capacitance.

In linear (triode) region the channel exists continuously from source to drain. Therefore, gate capacitance is usually divided into equal parts for source and drain regions [1]. In other words:

$$C_{gs} = C_{gd} = \frac{1}{2}WLC_{ox} \tag{6-16}$$

Unlike linear region, in saturation region the channel pinches off before reaching the drain region. In other words, drain voltage has a very small impact on either channel or gate charge which corresponds to $C_{gd} \approx 0$. It has been shown that, under these conditions, the gate-source capacitance is calculated from the following equation [1]:

$$C_{gs} = \frac{2}{3} WLC_{ox} \tag{6-17}$$

As discussed in Chapter 3, as an ESD protection device, MOSFET is used in different bias conditions, i.e. grounded gate, gate-coupled, substrate-triggered and gate-substrate-triggered configurations. In all MOS-based protection circuits drain is connected to the pad. However, gate connection is in such a way that the transistor is "off" under normal operating conditions. Hence, in order to calculate the parasitic capacitance of the MOS-based protection circuit, the small signal model in weak inversion region should be considered. As in all MOS-based protection circuits drain is connected to the pad, parasitic capacitance consists of the gate-drain capacitance in parallel with the drain-bulk capacitance. Based on equation (6-14), gate-drain capacitance is only the overlap capacitance, which is not a function of drain and source voltages. This feature makes this capacitor a linear capacitor which is not responsible for degradation in linearity of analog circuits. On the other hand, drain-bulk capacitance is a reverse-bias junction capacitance which is very non-linear. As a result, parasitic capacitance of MOS-based ESD protection circuits consists of a linear capacitance in parallel with a non-linear capacitance. Similar discussion holds for LVTSCR-based ESD protection circuits as well. In these devices, parasitic capacitance consists of a gate overlap capacitance in parallel with reverse-biased junction capacitance.

3. A 12-BIT 20 MS/S ANALOG TO DIGITAL CONVERTER [3]

As mentioned earlier, ESD protection circuits add highly non-linear parasitic capacitance to the main circuit. Non-linearity of the parasitic capacitance is very critical in analog to digital converters, where very high linearity at moderate and high frequency inputs is required. For instance, in wireless base stations that digitize the intermediate frequency band a spurious free dynamic range (SFDR) greater than 80 dB at several tens of MHz is required [4]. In one of the first studies I. E. Opris studied the interaction between the zener diode-based ESD protection and a 12 bit 20 MS/s A/D converter [3].

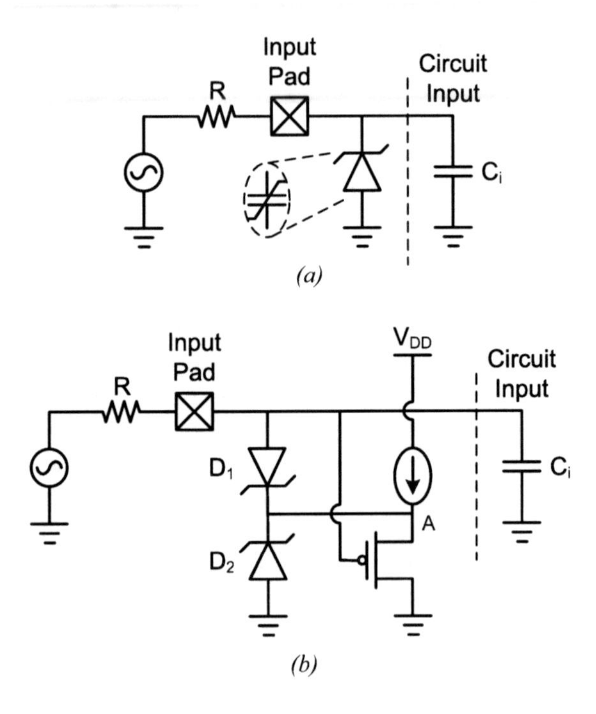

Figure 6-3. Providing ESD protection (a) with zener diode (b) with bootstrapped structure.

Figure 6-3(a) shows the equivalent circuit of the input circuit where the zener diode is modeled by a non-linear capacitor.

The junction capacitance of the zener diode is between 2 and 4 pF based on the input bias voltage. Resistor R equals to 25 Ω to model the 50 Ω resistance of the signal generator and 50 Ω termination resistance. Simulation results show that this ESD protection circuit has very high non-linearity and the second harmonic distortion is –62.3 dB [3]. The degradation in linearity is due to non-linear capacitance of the zener diode.

In order to improve the linearity of the ESD protection circuit, zener diode is replaced with two zener diodes in series as shown in Figure 6-3(b). It can be seen that, under ESD conditions, the protection mechanism hasn't changed as one of the diodes is forward biased and the other one is operating in zener breakdown mode. However, in normal operating conditions, node A is driven by a buffer, which consists of a PMOS source follower. Hence, the voltage across diodes has less variation and the value of non-linear capacitances doesn't change significantly. Therefore, linearity is improved. Moreover, the capacitance of D_1 is bootstrapped and as a result, its influence at the input pad is reduced. The bias current was designed to be 300 μA to provide large enough bandwidth and slew rate.

Both ESD protection circuits shown in Figure 6-3 were used to protect the 12-bit 20 MS/s analog to digital converter. Both circuits were implemented in 0.8 μm CMOS process. Figure 6-4 shows the spectrum of the input signal for the simple diode protection shown in Figure 6-3(a), while the spectrum of the input signal for the bootstrapped structure of Figure 6-3(b) is shown in Figure 6-5 [3].

It can be seen that, as suggested from simulation results, for conventional zener diode protection, the second harmonic is –56.7 dBV and linearity is very poor, while the third harmonic distortion is –67.1 dBV. On the other hand, bootstrapped ESD protection has much better linearity where the second harmonic is guaranteed to be better than –80 dBV.

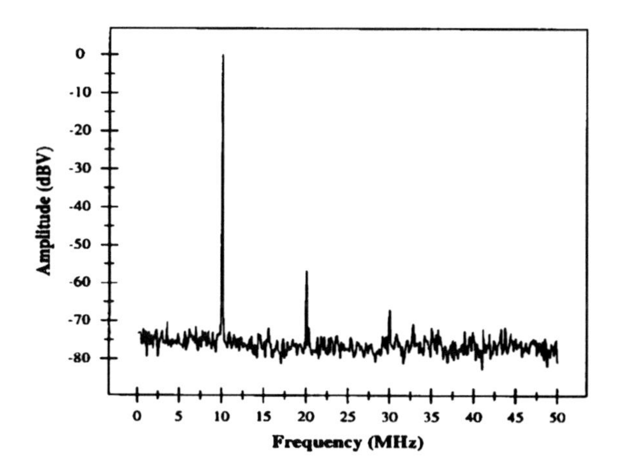

Figure 6-4. Spectrum of the input signal with zener diode ESD protection.

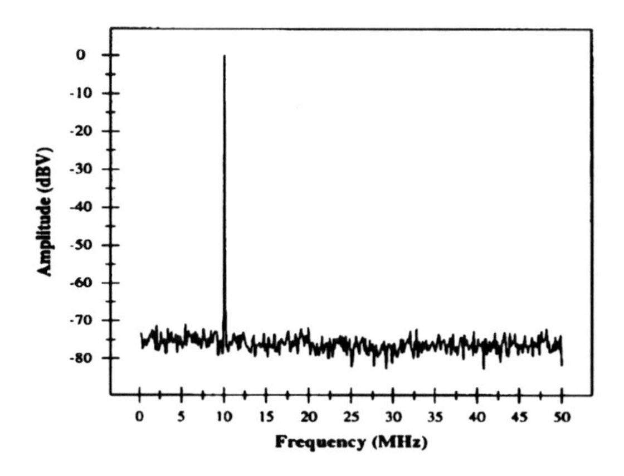

Figure 6-5. Spectrum of the input signal with bootstrapped ESD protection.

4. A 14-BIT 125 MS/S ANALOG TO DIGITAL CONVERTER [5]

The first case study in Section 3 showed that ESD protection circuit has great impact on linearity of analog to digital converters. As technology scales, even better performance will be required and hence, the distortion introduced by ESD protection circuits may become the limiting factor. Hence, it's necessary to fully understand the impact of ESD protection circuit on linearity. In this section, a model based on Volterra series analysis for second and third harmonic distortions is presented along with measurement results for both diode- and MOS-based protection circuits.

4.1 Volterra Series Analysis

In order to analyze the linearity of ESD protection circuits the circuit shown in Figure 6-6 is used [5].

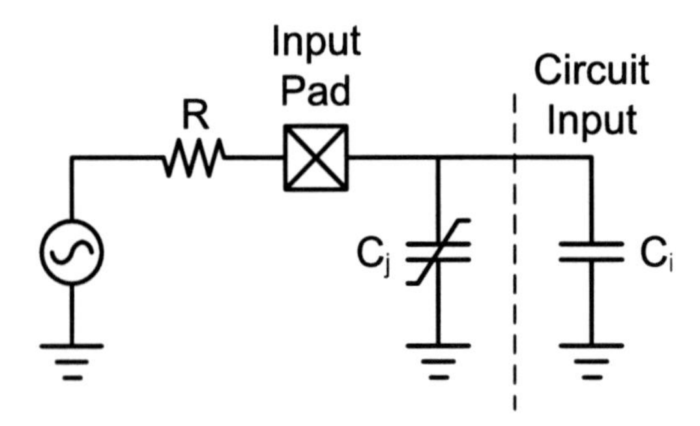

Figure 6-6. Setup to analyze the distortion caused by ESD protection circuits.

In this figure the main circuit is only modeled with its input capacitance C_i. ESD protection circuit is also modeled with its parasitic capacitance C_j. It should be noted that, as mentioned earlier, this capacitance mostly consists of reverse-biased junction capacitance, which is highly non-linear. In order to simplify the analysis, it is assumed that C_j is purely reverse-biased junction capacitance. Hence, the voltage dependence of this capacitor is given by equation (6-18) [1].

$$C_j = \frac{C_{j0}}{\left(1 + \frac{V_0}{\psi_0}\right)^M} \qquad (6\text{-}18)$$

In this equation ψ_0 is the *pn* junction built-in potential and M is the junction grading factor. In order to use this equation for linearity analysis, Taylor expansion is used for the right hand side of the equation. Equation (6-19) shows expanded equation with the first three terms of the Taylor expansion considered.

$$C_j = C_{jQ}\left[1 - M\frac{v_0}{V_Q + \psi_0} + \frac{1}{2}\left(M^2 + M\right).\left(\frac{v_0}{V_Q + \psi_0}\right)^2\right] \qquad (6\text{-}19)$$

In equation (6-19) Taylor expansion is written about the quiescent voltage V_Q, v_0 is the ac perturbance around this operating point and C_{jQ} is the junction capacitance at the quiescent point. It has been shown that using Volterra series analysis, the second and third order harmonic distortions of the output voltage (HD_2 and HD_3) are calculated from the following equations [4]:

$$HD_2(f_{in}) = \frac{1}{2}M\frac{f_{in}}{f_1}.\left|H(f_{in}).H(2f_{in})\right|.\left(\frac{v_i}{V_Q + \psi_0}\right) \qquad (6\text{-}20)$$

$$HD_3(f_{in}) \cong \frac{1}{8}\left(M^2 + M\right)\frac{f_{in}}{f_1}.\left|H(f_{in})^2.H(3f_{in})\right|.\left(\frac{v_i}{V_Q + \psi_0}\right)^2 \qquad (6\text{-}21)$$

In the above equations v_i is the peak input amplitude, while H(f), f_0 and f_1 are calculated from the following equations:

$$H(f) = \frac{1}{1 + j\frac{f}{f_0}} \qquad (6\text{-}22)$$

$$f_0 = \frac{1}{2\pi R\left(C_i + C_{jQ}\right)} \qquad (6\text{-}23)$$

$$f_1 = \frac{1}{2\pi R C_{jQ}} \qquad (6\text{-}24)$$

It should be noted that in equation (6-21) it is assumed that $f_{in} << f_1$ which is usually true in practice. Moreover, in order to avoid signal attenuation, the circuit is designed with f_{in} much smaller than f_0. As a result,

$|H(f_{in})|$, $|H(2\ f_{in})|$ and $|H(3\ f_{in})|$ are close to unity. Therefore, based on equations (6-20) and (6-21), low HD_2 and HD_3 are only possible if $f_l >> f_{in}$. From equation (6-24), C_{jQ} should be smaller which corresponds to smaller ESD device. Hence, there is a tradeoff between ESD robustness and the linearity.

The above model and discussions can be verified using an example. The circuit shown in Figure 6-3 is analyzed using the following parameters: M = 0.3, C_{j0} = 1 pF, C_i = 0, ψ_0 = 0.7 V, V_Q = 1.5 V, v_i = 0.5 V and R = 25 Ω. The value of R is based on assuming a 50 Ω input voltage source resistance and a 50 Ω termination resistance. Figure 6-7 shows HD_2 and HD_3 calculated from equations (6-20) and (6-21) and compares them with Spice simulation results [4].

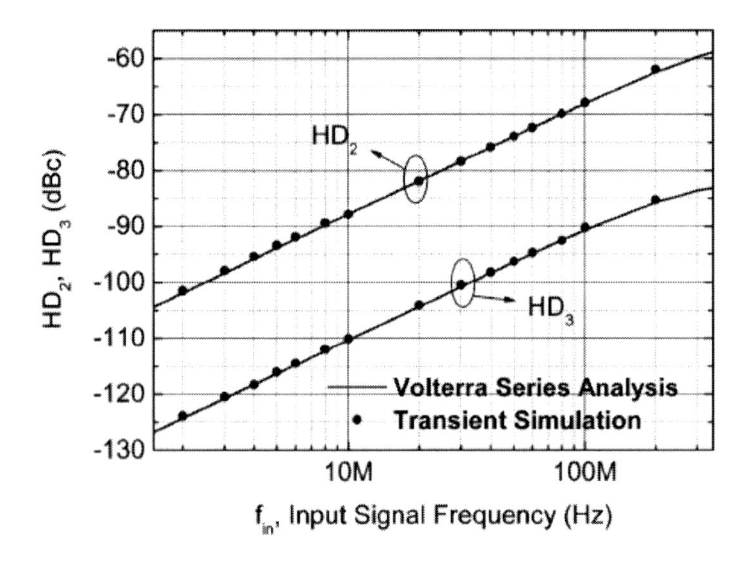

Figure 6-7. Verifying HD_2 and HD_3 equations with Spice simulations.

It can be seen that the model shows good agreement with Spice simulation results. It should be noted that HD_2 is often ignored in linearity analysis as it can be attenuated significantly with fully differential configurations. Hence, HD_3 calculated from equation (6-23) directly affects the linearity of the high performance circuits. Consequently, Figure 6-7 suggests that even by having a moderate junction capacitance, for 100 MHz performance HD_3 is limited to –90 dBc.

One of the main features of the analysis method is that it can be applied to any ESD protection circuit. This feature is due to the fact that reverse-biased junction capacitance is the main source of non-linearity. Even in MOS-based protection methods gate capacitance has a small variation, which makes C_{db} as the dominant source of non-linearity.

In the next section, diode and MOS-based ESD protection is provided for two analog to digital converters and the results are compared.

4.2 ADC with ESD Protection

In order to study the impact of ESD protection circuit on linearity of analog to digital converters, two separate devices were fabricated and tested. Device 1 is a 12 bit 65 MS/s ADC in 0.35 μm CMOS technology with SFDR of 85 dB at 32.5 MHz [6]. ESD protection for this device is provided using 100 × 77 μm dual diode as shown in Figure 6-8(a). Device 2 is a 14 bit 125 MS/s ADC in 0.18 μm CMOS technology with SFDR of 82 dB at 100 MHz [7]. ESD protection for this device is provided using gate-coupled NMOS as shown in Figure 6-8(b) [8]. Total area of this protection device is 90 × 221 μm. It should be noted that in order to do a fair comparison, both diode- and MOS-based protection devices were sized in such a way that the parasitic capacitance of each of them is approximately 0.7 pF.

Both of these devices were fabricated and their third harmonic distortion was measured. In order to be able to measure HD_3 of ESD protection

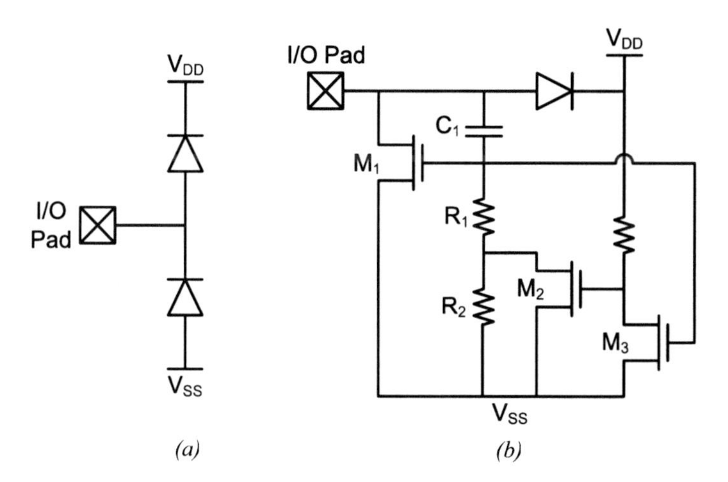

Figure 6-8. ESD protection devices used in this section (a) diode-based protection (b) MOS-based protection.

devices, a switch was added to ADCs that allowed disconnecting them from ESD protection circuit (holding mode). Hence, the HD_3 measurement of the ESD protection devices was done by placing the ADCs into hold mode. Figure 6-9 shows the measured third harmonic distortion of both ESD protection devices as a function of the input frequency [5].

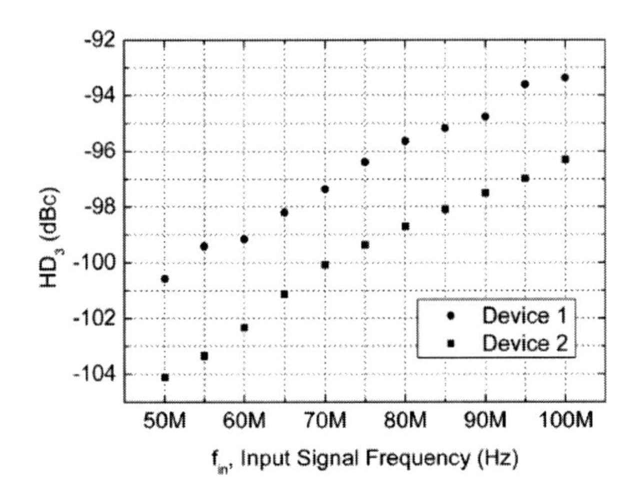

Figure 6-9. Comparing HD_3 of diode- and MOS-based ESD protection.

It can be seen that for both device measurements the value of HD_3 is below the ADCs' overall performance requirements. Moreover, although diode- and MOS-based protection circuits have similar parasitic capacitance, MOS-based protection (device 2) shows better linearity compared to diode-based protection (device 1). In order to understand the difference between linearity of diode and MOSFET, different components of their parasitic capacitance should be discussed. In diode, all the parasitic capacitance is reverse-biased junction capacitance, which is highly non-linear and corresponds to degraded linearity. On the other hand, in MOSFET, 25% of the capacitance is the gate capacitance which is almost constant above 1 V and the rest is reverse bias junction capacitance. In other words, MOSFET has a more linear capacitance over the input signal range of the ADC. Hence, as can be seen in Figure 6-9, MOS-based protection shows around 3 dB advantage in SFDR, even though both protection circuits have similar parasitic capacitance.

5. A 4 GB/S CURRENT MODE LOGIC DRIVER [9]

The growth of data transfer rate in telecommunication networks necessitates the design of high-speed circuits. Furthermore, scaling of CMOS technology into nanometer regime has enabled the full CMOS implementation of high speed integrated circuits. Buffers are one of the main blocks of high speed circuits such as clock and data recovery, serial to parallel converters and multiplexers/demultiplexers. The simplest implementation of a buffer consists of CMOS inverters. Figure 6-10 shows the schematic of a CMOS inverter along with its DC $V_{out} - V_{in}$ characteristic.

In addition to its simplicity, this buffer has a number of other advantages as well: It has a very low leakage current; as it can be concluded form Figure 6-8(b), this buffer has a very large small signal gain compared to other one stage buffers; finally, it has a very large noise margin. However, there are some major limitations in this buffer which makes it less suitable for very high-speed applications. The existence of a PMOS transistor lowers the maximum frequency of operation. Furthermore, the inverter is a single ended circuit which makes it very sensitive to noise sources such as supply noise, substrate noise and cross talk. Therefore, due to the above limitations, this buffer is not preferred in very high speed applications [10].

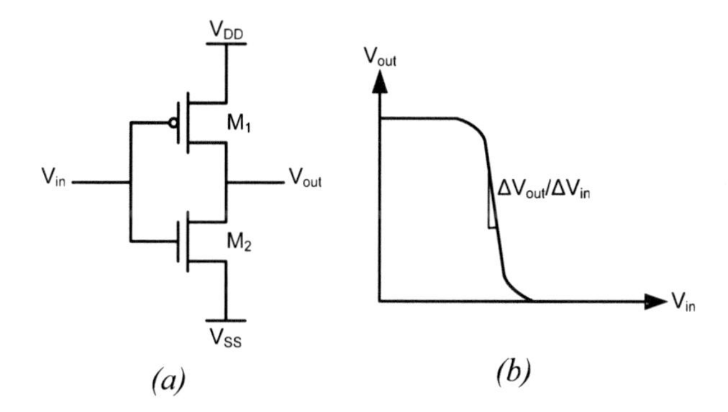

Figure 6-10. CMOS inverter (a) schematic (b) $V_{out} - V_{in}$ characteristic.

In high frequency applications where CMOS inverters cannot be used as buffers, current mode logic (CML) drivers are often used [10]. Current mode logic circuit was first used in gigahertz MOS adaptive pipeline technique [11]. However, due to its superior performance, it's been used in many other applications such as ultra-high-speed buffers [12], latches [13] and frequency dividers [14]. The main advantage of a CML circuit is the ability to operate with lower signal voltage and higher frequency at lower supply voltage. Furthermore, due to its differential structure and high common-mode rejection, CML buffer is insensitive to noises on power and ground nodes. Although they suffer from static power dissipation, due to their superior performance, they are the best choice for high speed applications [10]. Hence, in order to study the impact of different ESD protection methods on the behavior of high-speed circuits, a 4 Gbps CML driver is used as a design reference. ESD protection for this driver is provided for all four zapping modes and based on both MOS and SCR devices.

5.1 CML Driver Design

In order to see the impact of ESD protection circuits on high-speed drivers, a 4 Gb/s two stage CML driver was designed in 0.13 μm CMOS process. Differential output swing of the driver was targeted to be at least 800 mV with rise/fall time of less than 150 ps. At the same time, jitter should be less than 1 ps.

A two stage CML driver with 50 Ω on-chip loads was designed with above mentioned specifications. Figure 6-11 shows the schematic of the driver. The 50 Ω resistors R_{L1} and R_{L2} were designed using poly resistors. An of-chip bias resistor was also designed to facilitate the tuning of the output swing. Simulation results show an output swing of 850 mV with 116 ps rise time at 4 Gb/s data rate. A 4 Gb/s pseudo random input sequence is applied to the driver to simulate the jitter of the driver. The rms jitter is calculated from eye diagram drawn for 2,000 samples. The driver in Figure 6-9 shows a jitter of 229 fs.

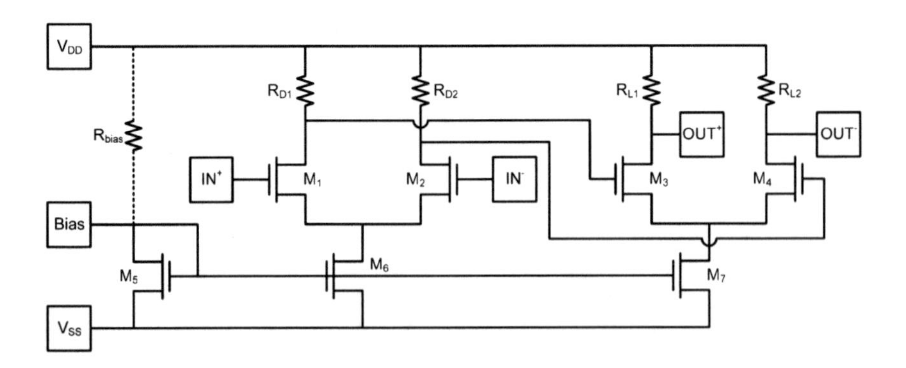

Figure 6-11. Two stage CML driver.

5.2 ESD Protection Methods

MOSFET and SCR based ESD protection devices were used in two different instances of the CML driver to provide complete ESD protection against all four zapping modes. As the circuit has five I/O and two supply pins, five ESD protection circuits are required for the I/O pins and one clamp between V_{DD} and V_{SS}. Figure 6-12 shows the protection scheme for this driver.

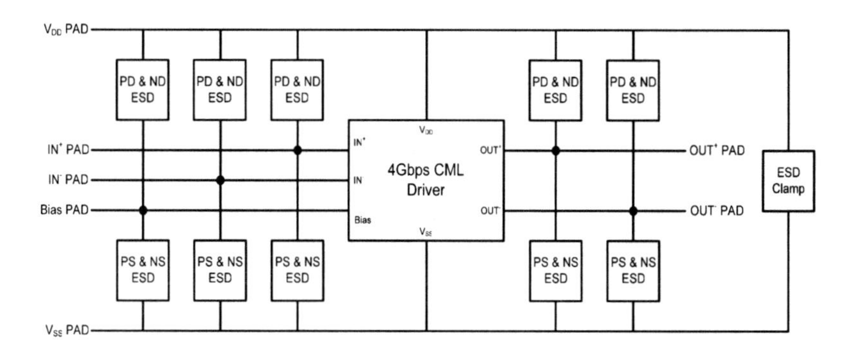

Figure 6-12. ESD protection scheme for the CML driver.

As discussed in Chapter 3, MOS and SCR are among the most popular ESD protection devices. However, due to their high first breakdown voltage, they are usually used with either gate-coupling [15, 16], or substrate triggering [17, 18], or gate-substrate triggering techniques [19]. To design a proper ESD protection circuit for the driver, the impact of different triggering techniques on the jitter of the driver is studied in this section.

5.2.1 Comparing MOSFET with SCR

The first instance of the driver is protected using 300 μm wide grounded-gate MOS transistors. Using *ac* simulation the total parasitic capacitance added to each pad is determined to be 663 fF. A 3 Gbps pseudo random input voltage is used to simulate the driver jitter. The output *rms* jitter is 1.27 ps. In MOS transistors the capacitance consists of both the drain-substrate junction capacitor and the gate-drain capacitor. On the other hand, in SCR device the capacitance mainly consists of well-substrate junction capacitance. To compare the impact of SCR capacitance with MOS capacitance, SCR is sized to have the same capacitance as the grounded-gate MOS and jitter is simulated. Simulation results show that jitter is the same for both SCR- and MOS-based protection. However, as SCR has a much higher ESD protection level per unit area the jitter of SCR-based protection is expected to be much lower than MOS-based protection.

5.2.2 Impact of Gate-Coupling on Jitter

As discussed in Chapter 3, gate triggering can be provided using a small transistor as shown in Figure 3-22. The most important benefit of this technique is the ability to reduce the first breakdown voltage with very small increase in the pin capacitance. To simulate the impact of the gate-triggering technique on jitter, a 5 μm transistor is added to the 300 μm ESD protection transistor. Due to small gate coupling transistor, the gate-coupled MOS protection has the same parasitic capacitance and protection level as the grounded-gate MOS protection. However, simulation results show that the *rms* jitter is increased from 1.27 to 6 ps. This significant degradation in jitter is due to the floating gate of the main ESD protection MOS transistor. Hence, another transistor is added to the gate-triggered MOS to set the gate-source voltage of the main protection device to ground potential under normal operating conditions. Figure 6-13 shows the schematic of the modified gate-triggered MOS protection.

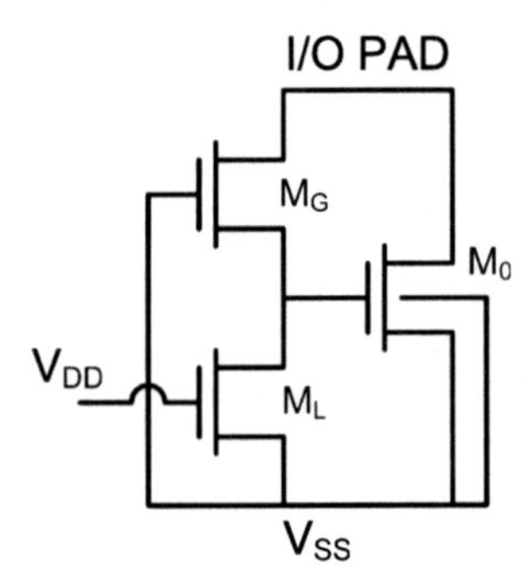

Figure 6-13. Modified gate-triggered NMOS.

In this figure, Mo is the main protection transistor which is 300 μm wide, MG is the gate-triggering transistor which is 5 μm wide, and ML is the extra transistor added to reduce the jitter. Under normal operating conditions ML turns on reducing leakage and improving jitter. On the other hand, under ESD conditions VDD is floating and the protection circuit is the same as the conventional gate-triggered NMOS.

Protection the CML driver with the new gate-coupled NMOS reduces the jitter from 6 to 1.8 ps. It should be noted that this device has similar first breakdown voltage and capacitance as the gate-triggered NMOS.

5.2.3 Impact of Substrate Triggering on Jitter

The impact of substrate-triggering on jitter is simulated by adding substrate triggering to the MOSFET as shown in Figure 3-25. Similar to the previous section, a 300 μm transistor is used as the main ESD protection transistor and substrate triggering is provided using a 100 μm transistor. Substrate-triggering has higher pin capacitance than grounded-gate MOS since the triggering transistor size is larger, while the protection level remains the same. Simulation results show that the *rms* jitter of the CML with substrate-triggered MOS protection is increased to 1.55 ps. Based on the mentioned simulation results it can be concluded that compared to gate-coupled MOS, substrate-triggered MOS has much smaller impact on performance of the CML driver.

5.3 CML Driver with MOS-Based ESD Protection

To design a MOS-based protection with low first breakdown voltage, gate-substrate triggering technique introduced in Chapter 3 is used. However, due to high leakage current and jitter of this structure, low leakage option is added to the gate-substrate triggered technique (refer to Figure 3-28). The ESD protection for PS and NS modes is provided using NMOS transistors. To provide the protection for PD and ND modes PMOS transistors are used. Figure 6-14 shows the schematic of the full mode protection that is used to protect the CML driver.

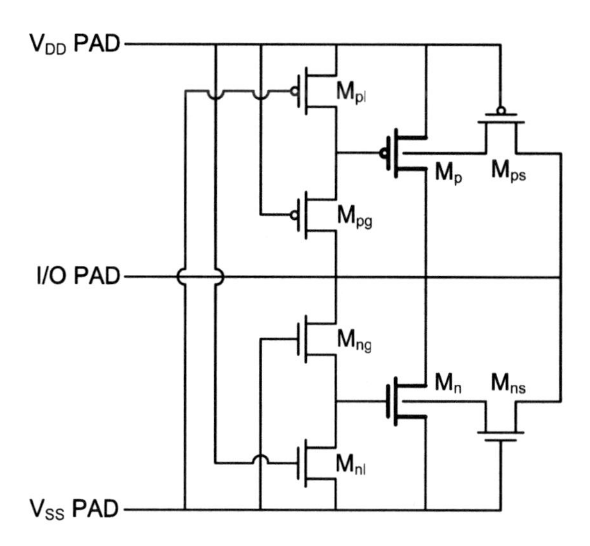

Figure 6-14. Full-mode ESD protection using modified gate-substrate triggered NMOS.

The main protection transistors are M_n and M_p which are 320 μm wide. Substrate triggering is provided with M_{ns} and M_{ps} and their width is 80 μm. Gate triggering is provided with 5 μm wide M_{ng} and M_{pg}. Finally, the leakage reduction and jitter improvement is done with transistors M_{nl} and M_{pl}. The driver shown in Figure 6-11 was implemented with ESD protection shown in Figure 6-14 in 0.13 μm CMOS process.

Several measurements were carried out to examine the effectiveness of the ESD protection circuits as well as their impact on circuit performance. First, the complete circuit is tested under ESD conditions. TLP measurement results show that the first breakdown voltage is 4.5 V and the second breakdown current is 1.76 A. This first breakdown voltage is low enough to provide protection in 0.13 μm CMOS process. Furthermore, HBM test show that ESD protection circuit passes ±3 kV stress.

Once the effectiveness of ESD structure was evaluated, the performance of the driver with MOS-based protection is evaluated. The pulse generator with an *rms* jitter of 7 ps was used for measurement. We should subtract the pulse generator induced jitter from the measured *rms* jitter at the driver's output. In this fashion, we can examine the impact of the driver and ESD protection circuit parasitics on the jitter. Figure 6-15 shows the single ended output waveform of the driver. It can be seen that the MOS-based protection circuit has the differential output swing of 500 mV and the rise time of 315 ps.

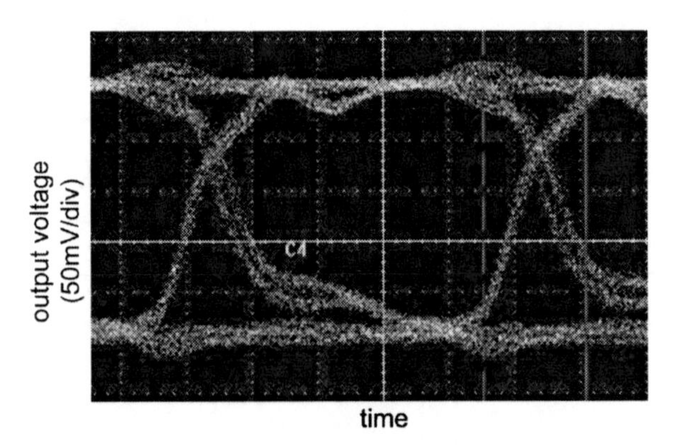

Figure 6-15. Output voltage of the driver with MOS-based ESD protection.

The jitter is measured by applying a 4 Gbps PRBS data sequence to the input of the driver. Figure 6-16 shows the eye-diagram of the output voltage for 2,000 samples. It can be seen that the output *rms* jitter is 10.7 ps. Hence, the driver with MOS-based ESD protection adds 3.7 ps to the input jitter.

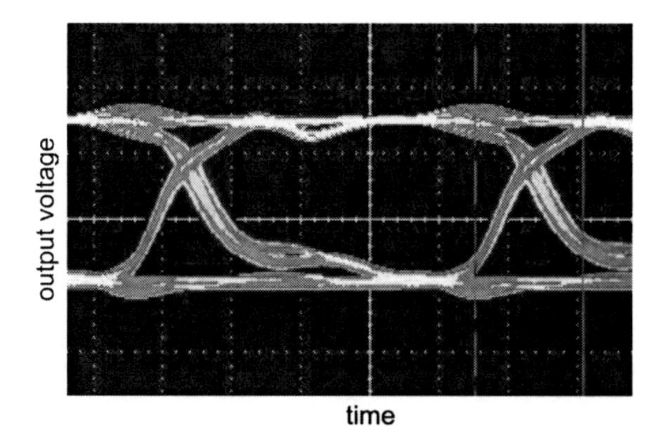

Figure 6-16. Measured eye-diagram of the driver with MOS-based ESD protection.

It can be seen that using a MOSFET-based protection, the driver performance, i.e. swing, rise/fall time and jitter, is degraded significantly.

5.4 CML Driver with SCR-Based ESD Protection

In order to provide ESD protection for the driver with SCR-based devices, LVTSCR with gate-substrate triggering is used. The ESD protection for all four zapping modes is provided based on the concept of all direction SCR discussed in Chapter 3. The cross section of the gate-substrate-triggered LVTSCR used to protect the CML driver is shown in Figure 6-17.

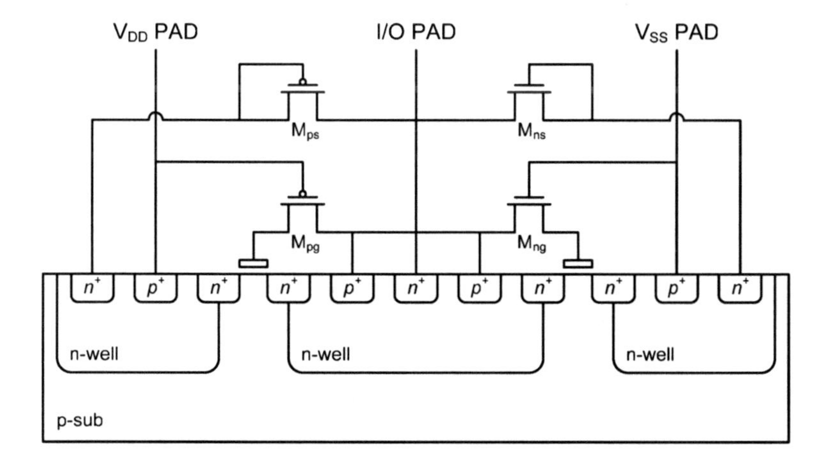

Figure 6-17. Cross section of the all-direction gate-substrate triggered LVTSCR.

In this figure M_{ng} and M_{pg} are used for gate-coupling, and M_{ns} and M_{ps} are used for substrate triggering. The SCR device is 100 μm wide. Substrate triggering MOSFETs are 20 μm wide while gate triggering MOSFETs are 5 μm wide.

Similar to Section 5.2, the ESD protection circuit of Figure 6-17 is added to all I/O pins of the CML driver of Figure 6-11 in 0.13 μm CMOS process. First, ESD response of the complete circuit is evaluated using TLP and HBM measurements. Measurement results show that the first breakdown voltage of the ESD protection circuit is 5 V and it passes ±3 kV HBM stresses.

Subsequently, performance of the driver with SCR-based ESD protection is evaluated for its design parameters. Similar to the previous section the pulse generator had an output *rms* jitter of 7 ps. Figure 6-18 shows the single-ended output voltage of the driver. It can be seen that the differential output swing is 700 mV and the rise time is 148 ps.

Finally, jitter is measured by applying a PRBS sequence of 4 Gbps data rate to the input of the driver. Figure 6-19 shows the eye-diagram of the output voltage for 2,000 samples. It can be seen that the jitter is 7.7 ps. Hence, the driver with SCR-based protection adds 700 fs to the input jitter.

Table 6-1 compares simulation results for the CML driver without ESD protection with measurement results for the driver with MOS-based and SCR-based protection methods. It can be seen that degradation in performance for SCR-based protection is much smaller compared to MOS-based protection. As that both ESD protection circuits have similar ESD

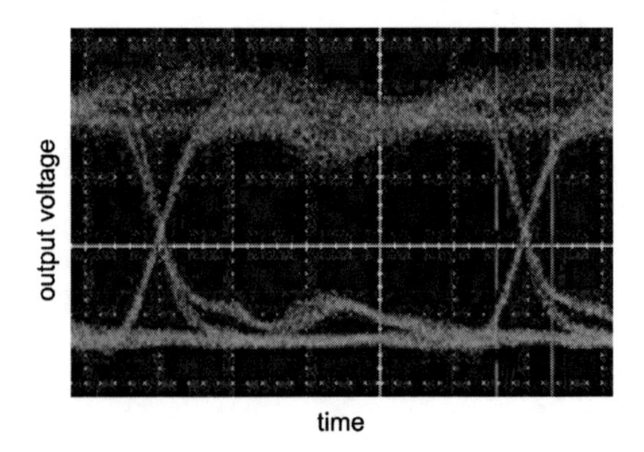

Figure 6-18. Output voltage of the driver with SCR-based ESD protection.

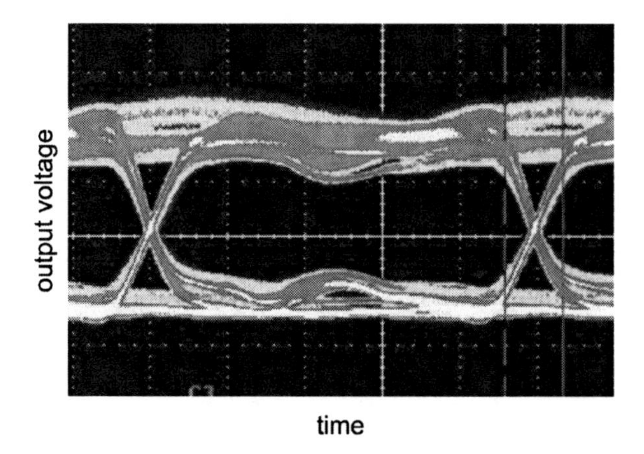

Figure 6-19. Measured eye-diagram of the driver with SCR-based ESD protection.

Table 6-1. Comparing different ESD protection techniques.

	Diff. output swing (mV)	Rise time (ps)	Input jitter (ps)	Output jitter (ps)
CML driver (simulation)	850	116	0	229
CML + MOS prot. (measurement)	500	315	7	10.7
CML + SCR prot. (measurement)	700	148	7	7.7

protection level of 3 kV, SCR-based is the best method for high-speed applications. This is generally due to higher ESD protection level per unit area for SCR-based circuits which allows higher ESD protection level with smaller parasitic capacitance.

5.5 Discussion on Jitter-Capacitance Relation

It can be seen from the ongoing discussion that ESD protection circuits, if not designed properly, can have significant impact on the I/O circuit performance. Based on the results in Sections 5.3 and 5.4, jitter is the most sensitive parameter to the parasitics. Therefore, minimizing the impact of ESD protection circuit on jitter is one of the important requirements in applications such as cable drivers. Therefore, it is useful to know the maximum allowed capacitance that can be added by ESD protection circuit without sacrificing the performance. In order to do this, ESD protection circuit is modeled with a capacitor and overall performance parameters are simulated using Cadence for different capacitor values. Table 6-2 shows maximum swing, rise time and jitter for different capacitor values.

Table 6-2. Impact of parasitic capacitance on driver performance.

Capacitance (fF)	Diff. output swing (mV)	Rise time (ps)	Output jitter (fs)
50	888	121	284
150	824	132	511
300	782	146	1,160
600	711	163	4,060

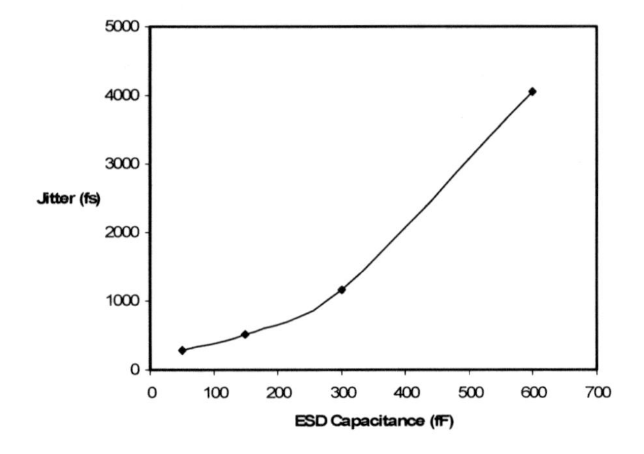

Figure 6-20. Impact of ESD capacitance on jitter.

It can be seen that, by changing the ESD capacitance from 50 to 600 fF, swing is reduced by only 15% while rise time is increased by 34% and *rms* jitter is increased by more than 14 times. Figure 6-20 shows the jitter of the driver as a function of the ESD capacitance. It can be seen that by increasing the capacitance over 150 fF the rate of increase in jitter increases significantly. Hence, this graph suggests that in applications where jitter is a requirement, such as cable drivers, the parasitic capacitance of the ESD protection circuit should be below 150 fF.

6. SUMMARY

In this chapter the impact of different ESD protection circuits on normal circuit behavior is discussed. This discussion was based on three different case studies. Through these case studies it was shown that ESD protection circuits degrade performance of the main circuit mainly by adding extra parasitic capacitance to I/O pads. As this parasitic capacitance has non-linear voltage dependency, it limits the operating frequency of the main circuit in addition to degrading linearity. The first two case studies in this chapter showed the impact of ESD capacitance on linearity of analog to digital converters. The results showed that, for the same parasitic capacitance, ESD MOSFET creates smaller degradation in linearity compared to *pn* junction based ESD protection. Finally, in the last section it was shown that reducing parasitic capacitance below 150 fF ensures acceptable jitter in CML drivers with gigabit per second data rates.

REFERENCES

[1] P. R. Gray, P. J. Hurst, S. H. Lewis, and R. G. Meyer, *Analysis and design of analog integrated circuits*, Wiley, New York, 2001.

[2] D. Johns and K. Margin, *Analog Integrated Circuit Design*, Wiley, New York, 1996.

[3] I. E. Opris, "Bootstrapped pad protection structure," *IEEE J. Solid-State Cir.*, vol. 33, No. 2, pp. 300–301, 1998.

[4] H. Pan, M. Segami, M. Choi, C. Ling, and A. A. Abidi, "A 3.3-V 12-b 50-MS/s A/D converter in 0.6 μm CMOS with over 80-dB SFDR," *IEEE J. Solid-State Cir.*, vol. 35, No. 12, pp. 1769–1780, 2000.

[5] J. H. Chun and B. Murmann, "Analysis and measurement of signal distortion due to ESD protection circuits," *IEEE J. Solid-State Cir.*, vol. 41, No. 10, pp. 2354–2358, 2006.

[6] 12-bit, 65 MSPS, 3.3 V analog-to-digital converter (ADS5221), Texas Instruments Inc, 2003.

[7] 14-bit 125 MSPS analog-to-digital converter (ADS5500), Texas Instruments Inc, 2003.

[8] A. Amerasekera and C. Duvvury, *ESD in silicon integrated circuits*, Wiley, New York, 2002.

[9] H. Sarbishaei, O. Semenov, and M. Sachdev, "Optimizing circuit performance and ESD protection for high-speed differential I/Os," *IEEE Custom. Int. Cir. Conf.*, pp. 149–152, 2007.

[10] P. Heydari , R. Mohanavelu, "Design of ultrahigh-speed low-voltage CMOS CML buffers and latches," *IEEE Trans. on VLSI Syst.*, vol. 12, No. 10, pp. 1081–1093, 2004.

[11] M. Mizuno, M. Yamashina, K. Furuta, H. Igura, H. Abiko, K. Okabe, A. Ono, and H. Yamada, "A GHz MOS adaptive pipeline technique using MOS current-mode logic," *IEEE J. Solid State Cir.*, vol. 31, No. 6, pp. 784–791, 1996.

[12] K. Iravani, F. Saleh, D. Lee, P. Fung, P. Ta, and G. Miller, "Clock and data recovery for 1.25 Gb/s Ethernet transceiver in 0.35 μm CMOS," *IEEE Custom. Int. Cir. Conf.*, pp. 261–264, 2001.

[13] H. T. Ng and D. A. Allstot, "CMOS current steering logic for low-voltage mixed-signal integrated circuits," *IEEE Trans. on VLSI Syst.*, vol. 5, No. 3, pp. 301–308, 1997.

[14] H. D. Wohlmuth, D. Kehrer, and W. Simburger, "A high sensitivity static 2:1 frequency divider up to 19 GHz in 120 nm CMOS," *IEEE Radio Freq. Int. Cir. (RFIC) Symp.*, pp. 231–234, 2002.

[15] M. D. Ker, C. Y. Wu, T. Cheng, and H. H. Chang, "Capacitor-couple ESD protection circuit for deep-submicron low-voltage CMOS ASIC," *IEEE Trans. VLSI Sys.*, vol. 4, No. 3, pp. 307–321, 1996.

[16] M. D. Ker, H. H. Chang, and C. Y. Wu, "A gate-coupled PTLSCR/NTLSCR ESD protection circuit for deep-submicron low-voltage CMOS IC's," *IEEE J. Solid State Cir.*, vol. 32, No. 1, pp. 38–51, 1997.

[17] M. D. Ker, T. Y. Chen, and C. Y. Wu, "ESD protection design in a 0.18 μm salicide CMOS technology by using substrate-triggered technique," *IEEE Int. Symp. Cir. Sys.*, vol. 4, pp. 754–757, 2001.

[18] M. D. Ker and K. C. Hsu, "Substrate-triggered SCR device for on-chip ESD protection in fully silicided sub-0.25 μm CMOS process," *IEEE Trans. Elec. Dev.*, vol. 50, No. 2, pp. 397–405, 2003.

[19] O. Semenov, H. Sarbishaei, V. Axelrad, and M. Sachdev, "Novel gate and substrate triggering techniques for deep sub-micron ESD protection devices," *Microelectronics Journal*, vol. 37, pp. 526–533, 2006.

Chapter 7

ESD PROTECTION FOR SMART POWER APPLICATIONS

1. INTRODUCTION

Often in compact microelectronic systems, it is necessary to combine different functional blocks on one chip to minimize the component count. These micro-electronic systems are realized as integrated circuits (ICs) using a variety of technologies. One such application includes mixed power (smart power) technology combining bipolar, CMOS and DMOS (Double-diffused MOS) devices in a single chip. Today mixed technologies are capable of integrating CMOS macro-cells as complex as microcomputer cores, power transistors and nonvolatile memories. Smart power approach enables designs with a mixture of CMOS logic, low voltage analogue and high voltage drive transistors on one chip. Typical applications of smart power technology are industrial power actuators and programmable logic controllers, camera application control systems, automobile electronics and liquid crystal display (LCD) driver ICs [1]. Since smart power ICs include low voltage devices ($V_{DD} \approx$ 1.8–5 V) and high voltage devices ($V_{DD} \approx$ 20–60 V), low/high voltage ESD protection devices should be integrated in the chip.

In this chapter, we will discuss ESD protection strategies of high voltage modules in smart power ICs. There are two general categories of ESD protection schemes used in I/Os: the non-self protecting scheme and the self protecting scheme. For non-self protecting scheme, it needs to add the ESD protection devices to the I/O cell. While for the self-protecting scheme, the I/O cell itself is an ESD protection device. In smart power technologies, the typical self protecting device is the lateral diffused MOS (LDMOS)

O. Semenov et al., ESD Protection Device and Circuit Design for Advanced CMOS Technologies, 173–197.

power transistor since the device can be used as the ESD protection device and as output driver simultaneously. However, since the parasitic lateral *npn* bi-polar transistor is difficult to be turned-on by the ESD pulse, the grounded-gate LDMOS often suffers the ESD vulnerability. The typical non-self protecting scheme for power technologies is based on ESD high-voltage MOSFETs, silicon controlled rectifier (SCR) devices and bipolar junction transistors (BJTs).

This chapter is focused on the analyzing of ESD robustness of different power ESD protection devices, latch-up immunity and layout issues.

2. LDMOS-BASED ESD PROTECTION

The lateral diffused MOS (LDMOS) transistor is a power transistor that is commonly used in high voltage circuit designs. The cross-section of typical LDMOS transistor is shown in Figure 7-1 [1]. The drain terminal is formed in N-well region and body and source terminals are formed in P-well region. In LDMOS transistor, the avalanche generation is initiated by the n^+ buried layer (NBL)/P-well junction. The gate oxide thickness and effective channel length are determined by the required operating voltage. For example, 500 Å of gate oxide thickness and 1 μm of effective channel length are typically used for 70 V applications [2]. The width of LDMOS ESD devices is varied in a wide range (200–2,400 μm) to pass 2–6 kV HBM ESD stress. The basic device structure can be optimized for different breakdown voltages (20 or 40 V) by the varying of N-well doping level [3].

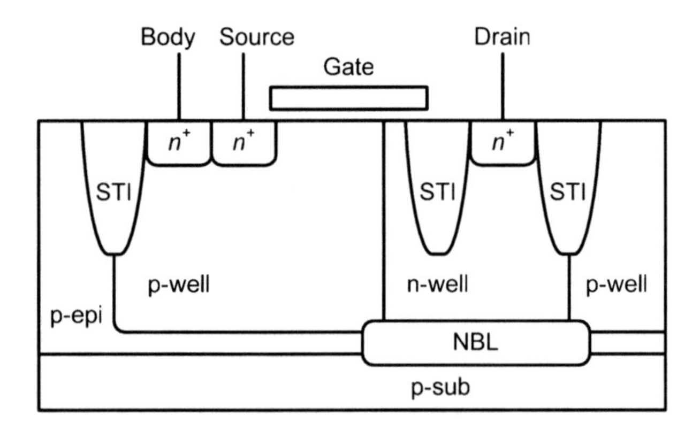

Figure 7-1. The cross-section of LDMOS power transistor.

The typical ESD protection design based on LDMOS transistor triggered by string of Zener diodes is shown in Figure 7-2. Zener clamps are used to prevent the device from entering bipolar breakdown when transient ESD peaks appear at the drain of LDMOS device.

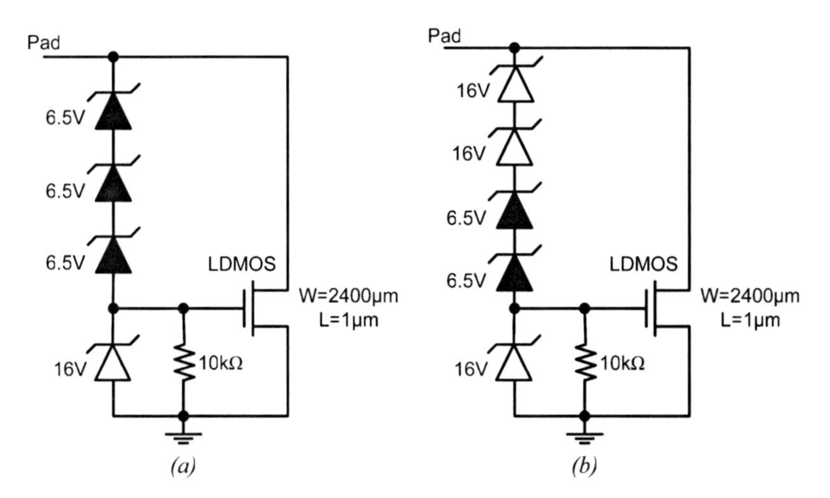

Figure 7-2. ESD protection design based on LDMOS transistor: (a) 2,400 μm/1 μm device with VDC = 19.5 V and VGC = 16 V, (b) 2,400 μm/1 μm device with VDC = 45 V and VGC = 16 V. (Adapted from [2].)

Under ESD event, the LDMOS operates as a saturated MOSFET and therefore a lower drain clamp voltage should result in better ESD performance. Thus it is expected that the LDMOS ESD performance will depend on the gate clamp voltage V_{GC}, drain clamp voltage V_{DC}, and the device width W. For large device widths with many multiple gates, uniform current flow is not easily obtained as this requires all the gates to be simultaneously turned on to the same potential. Inserting a small resistor in the path between the drain and the Zener stack should allow more uniform charging of the gate. Based on experimental results, the empirical relation between the ESD failure threshold and the gate/drain clamp voltages was derived [2]:

$$V_{ESD} = \frac{0.5\left(V_{GC} - V_{TH}\right)^2 \times W}{\left[1 + 0.1\left(V_{GC} - V_{TH}\right)\right] \times V_{DC}} \tag{7.1}$$

where W is the device width in μm, V_{GC} and V_{DC} are the gate and drain clamp voltages in volts, V_{TH} is the threshold voltage in volts, and V_{ESD} is the HBM level in volts. The constant 0.5 is the technology dependent parameter. To optimize the ESD performance of ESD protection design shown in

Figure 7-1, a 10 kΩ resistor can be inserted between the gate of LDMOS transistor and ground. The Zener diodes in the drain and gate clamps should be sufficiently large (>50 μm) to survive the ESD current before the LDMOS conduction begins.

In order to investigate the high current characteristics of LDMOS ESD transistor and to detect its triggering voltage (V_{tr}), holding voltage (V_h), on-resistance (R_{on}) and second breakdown current (I_{t2}), the TLP measurements are generally performed. The typical leakage and high current *I-V* characteristics of the gate grounded 40 V-LDMOS transistor are depicted in Figure 7-3 [4]. The obvious advantage of LDMOS based ESD protection circuits for the Smart Power application is that these devices can be implemented using standard CMOS process. However, LDMOS ESD transistors typically have a relatively low ESD robustness (2–4 KV) and low holding voltage (~5–8 V). Such ESD protection devices with low holding voltage in power-rail ESD clamp circuits may cause latchup failure in high voltage CMOS ICs.

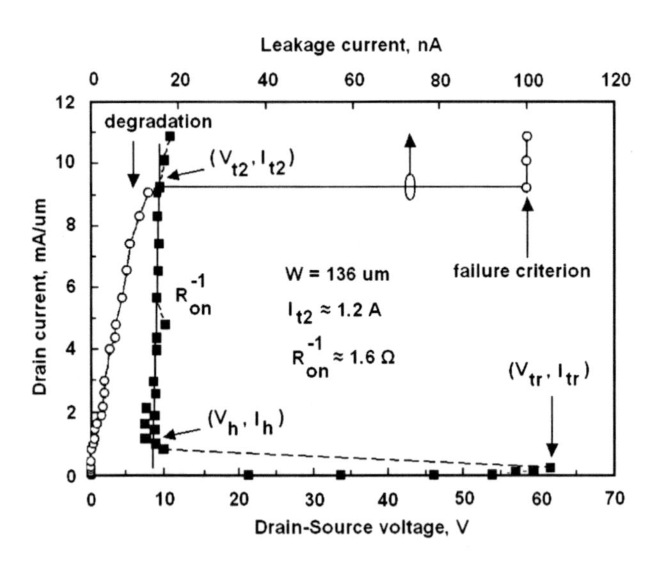

Figure 7-3. Typical TLP measurement data of 40 V-LDMOS transistor with grounded gate. (Adapted from [4].)

To increase the ESD robustness of LDMOS based protection circuits, the Darlington configuration of LDMOS transistors was developed as shown in Figure 7-4 [5]. This ESD circuit design was practically implemented using 20 and 40 V LDMOS devices. In this ESD circuit, the upper transistor (with

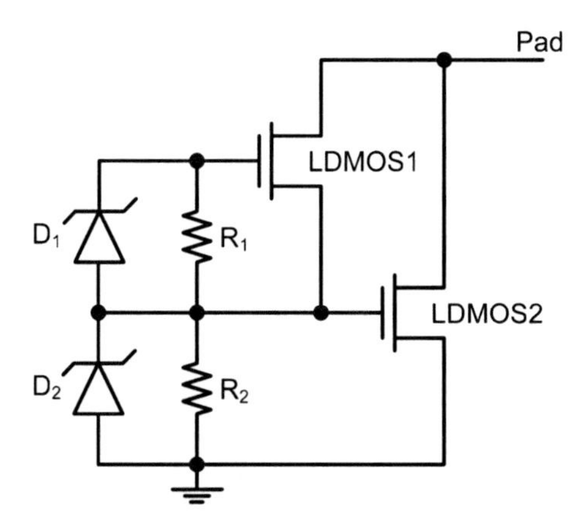

Figure 7-4. ESD protection circuit based on LDMOS transistors in Darlington configuration.

a minimum device width) turns on due to the capacitive coupling effect and it forces a current into the resistor R2. The lower LDMOS transistor (with a large width) turns on due to the capacitive coupling effect and due to the voltage drop across the resistor R2. The measured HBM ESD robustness of LDMOS based protection circuit in Darlington configuration was more than 8 kV for 20 V-LDMOS devices and it was around 5 kV for 40 V-LDMOS devices.

The failure analysis of ESD LDMOS transistor shows that the typical failure mode of this device is the drain-to-source filamentation due to a strong non-uniform triggering in a multi-finger transistor [4, 5]. It is known that the simultaneous triggering of multi-finger transistor can be achieved if the second breakdown voltage (V_{t2}) is larger than the triggering voltage (V_{tr}). Therefore, it is beneficial to increase V_{t2} or/and reduce V_{tr} in order to adjust V_{t2} above V_{tr}. Generally, a reduction of V_{tr} below V_{t2} can be accomplished by slightly increasing the substrate (bulk) potential or, alternatively, by applying a positive gate bias [6]. However, the recently published experiment-tal data shows that the above mentioned condition ($V_{t2} > V_{tr}$) can not be reached for the LDMOS device even for gate voltages up to 16 V [4]. As a consequence, biasing the gate does not improve the HBM robustness of any single and multi-finger device. Another option to improve homogeneous current flow is to increase the LDMOS on-resistance (R_{on}) and thus V_{t2} by an enlargement of the drain/source contact to gate spacing (DCGS/SCGS) in the non-silicided smart power process. However, for self-protecting output drivers, this means an increased MOS on-resistance and thus a reduced driver capability.

3. BJT-BASED ESD PROTECTION

The integration of power devices and of both analog and digital circuitry on the same chip makes it possible to have a fully integrated system with a significant increase in reliability and packaging capabilities. Smart-power technologies with junction isolation and vertical power current flow are particularly interesting since the vertical current flow allows to implement the power-transistors with less area and using the standard processes steps. At present, two modifications of smart power technologies are widely used. These are the Vertical Intelligent Power (VIPower) technology and the BCD (Bipolar + CMOS + DMOS) technology [7, 8]. The cross-section of power devices implemented in VIPower technology is shown in Figure 7-5. Since both of these technologies can integrate CMOS and bipolar devices in the same chip, the chip designers can use BJT and SCR based devices for ESD protection circuits.

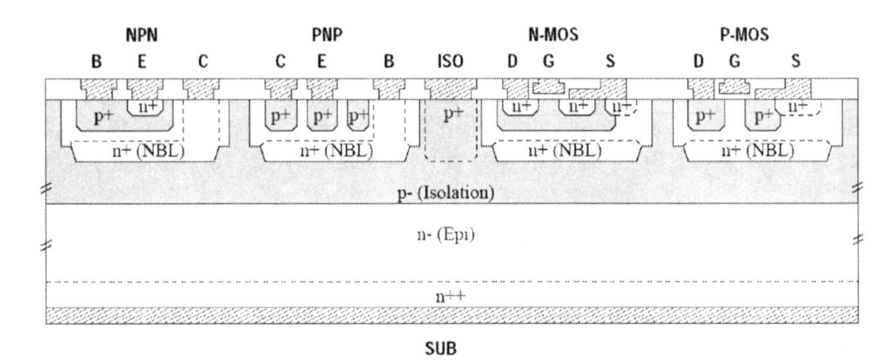

Figure 7-5. Cross-section of power devices implemented in VIPower technology.

One of the popular ESD protection devices in smart power technologies is the vertical grounded based bipolar *npn* transistor (VGBNPN) [9, 10]. Figure 7-6 shows the electrical schematic and the cross-section of this device. When a negative ESD stress is applied to the collector of the VGBNPN, it behaves as a forward-biased diode. In this mode the ESD robustness is high due to a relatively low electric field in the device. In case of positive ESD stress, the VGBNPN is triggered when the voltage on the collector electrode reaches the base-collector breakdown voltage. The resulting avalanche current flowing through the internal base resistance then forward-biases the base-emitter junction and triggers the *npn* bipolar transistor on. Once the *npn* transistor turns on, the avalanche and bipolar effects are combined resulting in the decrease of collector voltage down to the

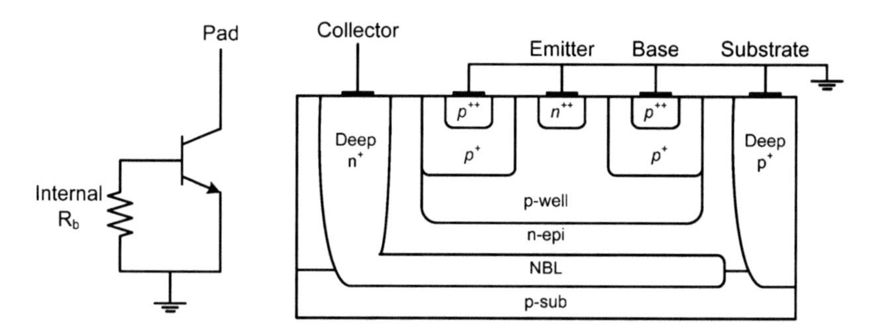

Figure 7-6. Electrical schematic and cross-section of the ESD VGBNPN transistor. (Adapted from [10].)

snapback holding voltage. This ESD structure provides a very good ESD robustness ranging from 3.5 kV (26 V/μm) to more than 10 kV (75 V/μm) under positive HBM stress and from 700 V (5 V/μm) to 1,300 V (9.5 V/μm) under positive MM stress [10]. To get information on the dynamic behavior of the VGBNPN transistor, such as triggering voltage and current, holding voltage and on-resistance, often the TLP measurements are performed.

The typical TLP characteristics of VGBNPN transistor implemented in a 1.8 μm 65 V smart power technology is depicted in Figure 7-7. This ESD device has 67 V of triggering voltage and 25 V of holding voltage. The 2-D TCAD simulations of VGBNPN transistor performed at 8 kV ESD stress are shown that the maximum electric field and the hot spot region are located below the emitter electrode at the N-epi/N-BL junction interface.

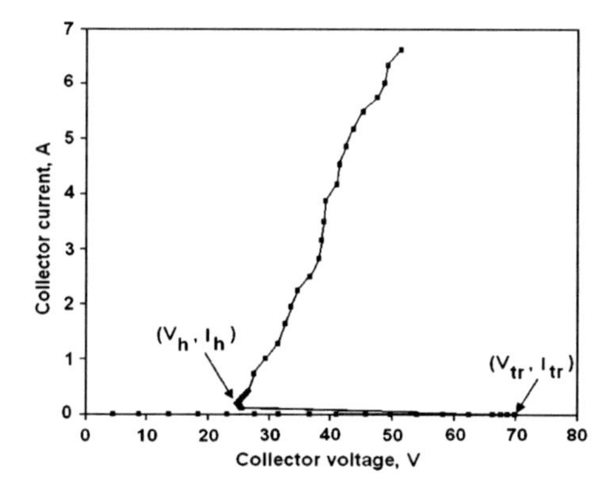

Figure 7-7. TLP measurement data for the VGBNPN ESD transistor. (Adapted from [10].)

Device simulation results were supported by the failure analysis carried out using the focused ion beam technique (FIB).The de-processing revealed that the melted silicon region due to the thermal breakdown is located deep below the emitter, at a location where 2-D simulations showed both high electric field and strong self-heating effect [10, 11].

The lateral *npn* BJT (bipolar junction transistor) power transistor can be also used for ESD protection circuits in smart power technologies. Since smart power ICs are frequently use a multiple V_{DD} design, the ESD protection circuits require ESD devices with high electrical flexibility (strong influence of the layout on the electrical properties) to meet the different electrical requirements. It was demonstrated that the holding/triggering voltage of protection BJT devices can be easily tuned by the change of layout parameters. The cross-section of high voltage BJT ESD device implemented in a 0.7 μm CMOS process is shown in Figure 7-8 [12].

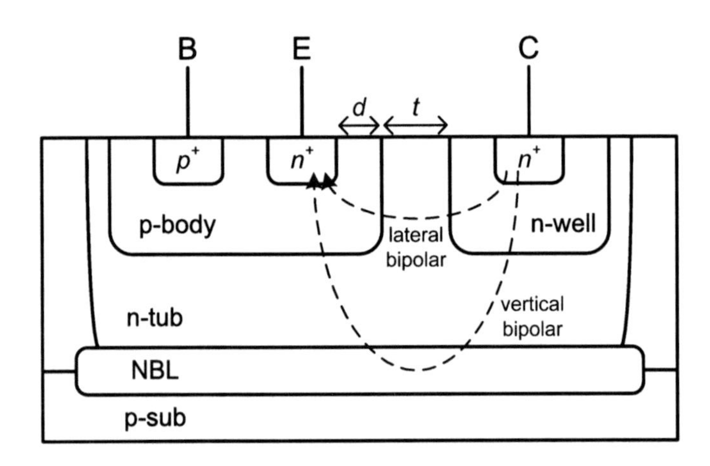

Figure 7-8. Cross-section of the npn ESD bipolar device and the current path of two bipolar structures triggered at ESD stress.

It consists of a *p*-substrate, a n^+ buried layer (BLN) and a low *n*-doped region called *n*-tub. The n^+ collector is in an *n*-well while the n^+ emitter and the p^+ base are both in a *p*-doped region called *p*-body. In order to control the ESD parameters, two main layout parameters are varied, as shown on Figure 7-8. It is the distance between the *p*-body and *n*-well (t parameter), and *d* is the distance between the *p*-body edge and the n^+ emitter. By the changing of parameter *t,* the triggering voltage (V_{tr}) can be varied due to the reach-through effect. The reach out effect is when the p-body/n-tub junction becomes closer to the highly doped *n*-well leading to a local increase of the

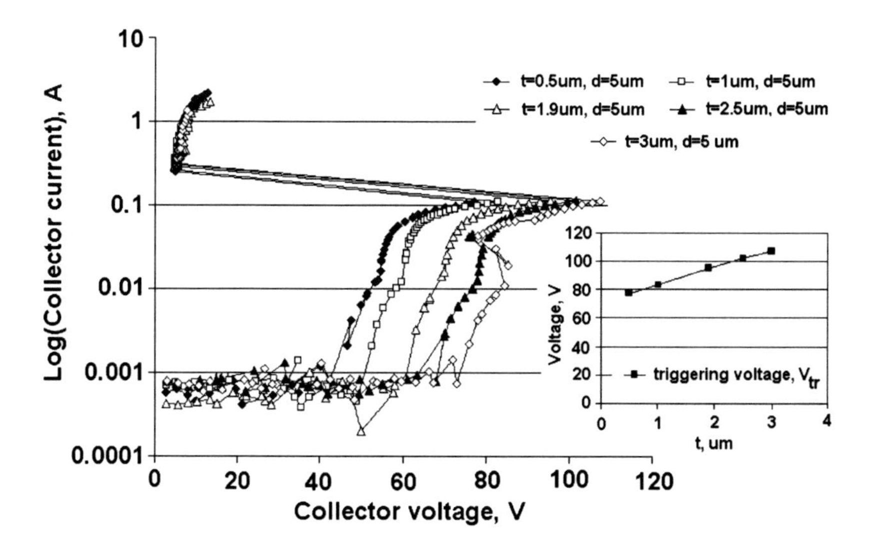

Figure 7-9. TLP IV characteristics and triggering voltage of npn ESD BJT transistor for different parameter "d". (Adapted from [12].)

electric field and earlier breakdown. The *d* parameter is aimed to control the holding voltage (V_h) by the changing of lateral base width.

The TLP measurements of devices with a constant *d* = 5 μm and *t* varying from 0.5 to 3 μm are depicted in Figure 7-9. These *IV* characteristics show that *t* influences strongly on the triggering voltage of the device. As *t* becomes large, the triggering voltage increases from 75 to 110 V. The TLP measurements of devices with a constant t = 4 μm and *d* varying from 3 to 20 μm are depicted in Figure 7-10. This figure shows that the holding voltage drops to 5 V for *d* = 3 μm to 40 V for *d* = 20 μm and changed gradually from 40 to 10 V for *d* = 10 μm with the current increase. The analysis of *I-V* characteristics obtained for the different *d* revealed that two bipolar structures can be triggered in this ESD device at ESD stress: a lateral one at the surface and a vertical one via the n^+ buried layer (BLN). This is sketched on Figure 7-8. 2-D TCAD simulations shown when "d" increases, the gain (β) of the lateral bipolar structure decreases and at a certain point it switches off and the vertical bipolar structure is triggered and the current is mainly flowing vertically via the BLN. Finally, by varying of the parameter *d*, the type (vertical or lateral) of the bipolar turn-on mechanism can be chosen. For the intermediate values of *d* (10 μm in this case), both lateral and vertical bipolar structures are in competition with each other. During the ESD stress the lateral bipolar structure is becoming more and more dominant and it allows the holding voltage to drop from the vertical bipolar holding voltage to the lateral bipolar holding voltage, as it shown on Figure 7-10.

The ESD performance (second breakdown current (I_{t2})) of devices with low/high holding voltage extracted from TLP measurements is presented in Figure 7-10. Due to low holding voltage, ESD device with d = 3 μm shows very robust capabilities (16 mA/μm). Together with the possibility to tune the trigger voltage using the "t" parameter, it makes this device suitable for I/O ESD protection. On the other hand, the ESD device with d = 20 μm has a large holding voltage (40 V). Hence, this device is suitable for power rails protection. It fails at 1.6 A/80 μm (20 mA/μm) [12]. At the intermediate d (=5 μm), the combination of a part of the current flowing laterally at the surface and a larger holding voltage makes the device weak. Its I_{t2} current drops up to ~6 mA/μm.

The modern trend of ESD design in interface, telecom, automobile and display applications is to require the high levels of on-chip protection. Typical system requirements such as the IEC1000-4-2 specification correspond to ESD voltages greater than 15 kV Human Body Model (HBM) levels, requiring >10 A peak current as measured by TLP [13]. The conventional approach to meet this requirement is to protect the circuit by *npn*

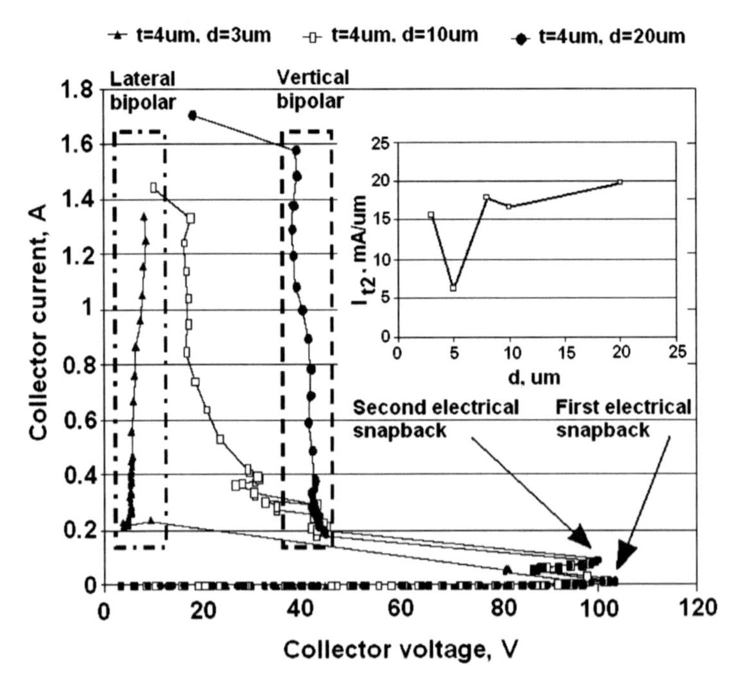

Figure 7-10. TLP IV characteristics and second breakdown current (I_{t2}) of npn ESD BJT transistor for different parameter "d". (Adapted from [12].)

bipolar structures. However, these devices are not very efficient due to the high level of power dissipation and unstable bipolar device operation [14].

4. SCR-BASED ESD PROTECTION

Silicon-controlled rectifiers use BJTs to conduct current and are used extensively in power device applications because of the capability to switch from a very high impedance state to a very low impedance state. Due to the low holding voltage, the power dissipation of SCR devices during an ESD event is typically less than that of other ESD protection devices. The SCR is the most efficient of all protection devices in terms of ESD performance per unit area. Hence, a properly designed SCR can be very attractive for ESD protection circuit in high voltage applications. Generally, SCR has a significantly lower peak of lattice temperature than BJT with the same silicon area at ESD stress, as shown in Figure 7-11. The cross-section of typical high-voltage SCR is depicted in Figure 7-12(a) [15].

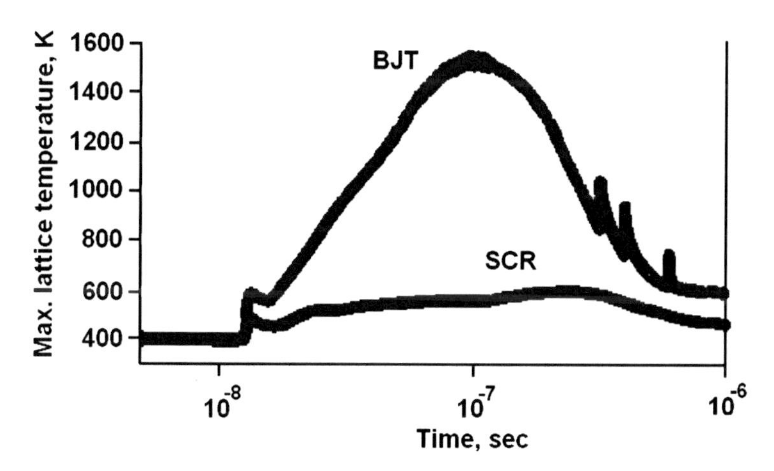

Figure 7-11. Peak temperature distribution in BJT and SCR at 5 kV HBM ESD stress (2-D TCAD simulations). (Adapted from [13].)

The TLP measured *I-V* characteristics of this ESD device, implemented in 0.25 μm 40 V CMOS process, is shown in Figure 7-12(b). The holding voltage of the SCR is ~4 V and the second breakdown current (I_{t2}) of the SCR with 200 μm width is over 6 A. Since the current flow in an SCR

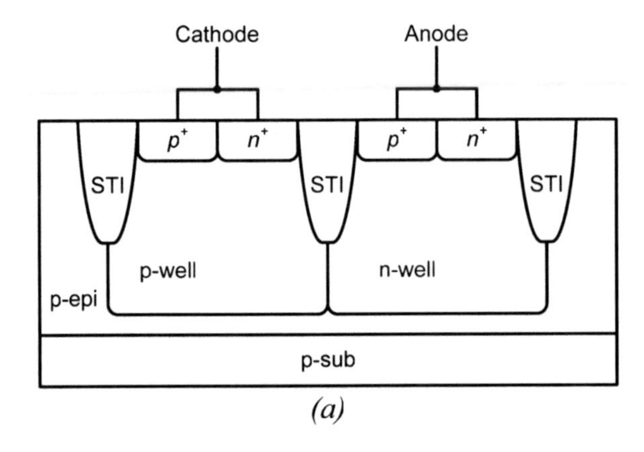

(a)

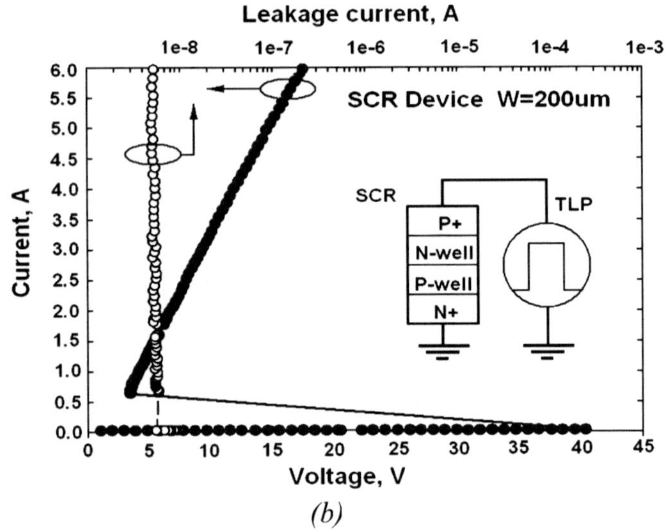

(b)

Figure 7-12. (a) Cross-section of high-voltage SCR, (b) TLP measured IV characteristics of high-voltage SCR. (Adapted from [15].)

device is always from its anode to its cathode, an SCR device can only be used for the single-polarity of ESD stress. When the ESD pulse has a positive voltage with respect to the V_{SS} node, the lateral SCR device will be triggered into its low impedance state to bypass the ESD energy through the SCR device to V_{SS}. But it cannot be triggered on if the ESD pulse has a negative voltage with respect to the V_{SS} node. For negative ESD protection, therefore, the parasitic junction diode formed by p-substrate and n-well in the structure of the lateral SCR device dissipates the ESD energy. This degrades the protection capability of overall ESD protection circuit.

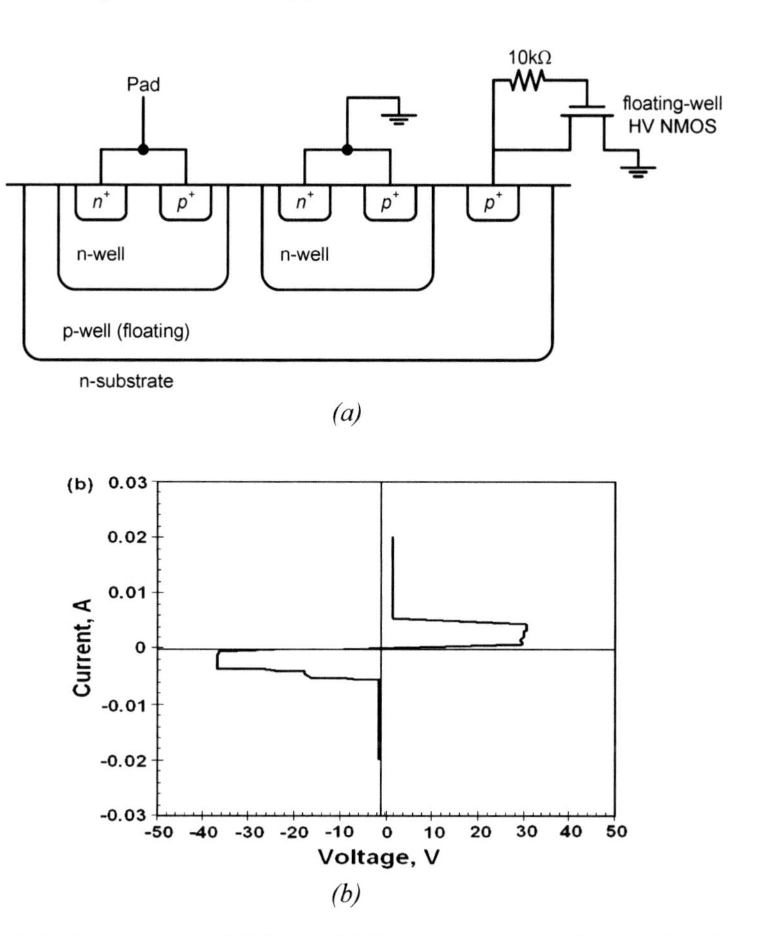

Figure 7-13. (a) Bi-directional SCR based ESD protection circuit for HV I/Os, (b) the IV curves of the bi-directional SCR. (Adapted from [17].)

To overcome this disadvantage of conventional SCR, the dual lateral SCR structure was developed, as shown in Figure 7-13(a) [16, 17]. In this protection circuit, a lateral SCR structure is arranged to discharge the positive as well as the negative ESD pulses. As an example of usage of dual lateral SCR structure for ESD protection, let's consider the ESD protection of SERIAL I/O IC's [17]. The SERIAL I/O is a CMOS Input/Output Transceiver IC for Personal Computer (PC) applications. It features a special clamping property to wake up PCs during the system standby mode. This chip was fabricated in 0.8 μm 30 V *n*-substrate process to implement high voltage functions between ±12 V. The design of I/O protection circuit in SERIAL I/O IC is one of the most challenging tasks. The high voltage (HV) input devices have to withstand the input signal from the external source when all IC power pins are shorted to ground. Therefore, the ESD protection of HV input circuit can not use any *pn*-junction directly connected to the input pad. Otherwise the ESD device will be forward biased and the input wake-up signal will be shorted to ground. This requires that all active devices directly connected to HV input pads must be located in floating wells to isolate input clock signal.

The ESD protection circuit, developed for HV input pads, is composed of a bi-directional SCR as a primary ESD protection device and a HV NMOS as a secondary protection device to trigger the bi-directional SCR, as shown in Figure 7-13(a). The bi-directional SCR device exhibits symmetrical IV shape and very good negative-resistance behavior (Figure 7-13(b)). Its ESD threshold level can be as high as 5 kV. Nevertheless,the response time of a standalone bidirectional SCR is too slow to protect internal circuit. Thus an HV NMOS connected to the floating well of the bi-directional SCR acts as an early triggering device to push bi-directional SCR into turn-on state during ESD events. In the ESD protection circuit shown in Figure 7-13(a), a triggering HV NMOS transistor with 10 kΩ resistor coupled to the gate is connected to the floating *p*-well of the bi-directional SCR cell. The measured HBM ESD robustness reached 5 kV, but latchup tests failed at 50 mA of triggering current [17]. The failure analysis found that if the HV NMOS transistor is separated from the floating *p*-well substrate of bi-directional SCR device using the FIB (focus ion beam) technique, the latchup problem can be solved, but the ESD robustness degrades up to 2 kV. Hence, the actual critical link is the connection of HV NMOS transistor to the SCR floating well. The latchup paths provide a conduction route during ESD event. Therefore, when latchup paths are removed latchup immunity improves, but the ESD protection level decreases. ESD and latchup are frequently two competing factors that must be taken into account together in order to improve the overall reliability and robustness of the design.

Finally, the bi-directional SCR is good enough to withstand the high ESD stress, but as mentioned above, it responds slowly with respect to ESD transient pulse waveform. It can not absorb ESD power promptly, and this power flows into internal circuits and leads to permanent damage due to its high switching voltage. The situation becomes worse for CDM ESD stress where the ESD transient has higher frequency content. Hence, the careful design of triggering schemes for protection circuits is essential to fully exploit the ability of bi-directional SCR ESD devices and is vital to its ESD robustness in products. Typical failure of bi-directional SCR structure is caused by the triggering of bi-directional SCR ESD cell that is induced by the HV N-MOSFET at the HV input. The reason is that HV N-MOSFET injects carriers into the bi-directional SCR floating p-well through triggering metal line during continuous ESD/latch-up stress. When the floating *p*-well has collected sufficient charge amount, the potential will drive carriers to trigger the bi-directional SCR into turn-on state. Hence, the HV N-MOSFET connecting to the floating well of the bi-directional SCR is the weakest link in the ESD protection circuit shown in Figure 7-13(a).

Another design of bi-direction SCR device named the Mirrored Lateral SCR (MILSCR) was developed for high voltage automobile applications [11]. In an automobile, a 12 V battery voltage may reach a value as high as +40 V when submitted to a load dump (disconnection of the battery from the alternator when the engine is running). In addition, to meet the inductive load discharge specifications, the clamping voltage for the inductive switching must be –40 V. Consequently, in normal operation, the output pin of automobile circuits should withstand –40/+40 V without activating the ESD protection structure. The ESD specification of this output pin is typically 2 kV HBM, but may be increased to 4 kV HBM and even 8 kV HBM according to the customers' requirements. To absorb such high ESD energy, the protection structure should have a low on-resistance and holding voltage [11]. In addition, protection structures dissipating their energy in some volume of silicon and not in a thin channel region as MOSFETs, are the most appropriate. Bipolar transistors offer this advantage but at the price of an inadequate on-resistance and holding voltage, whereas the silicon controlled rectifier combines both properties. The MILSCR ESD protection device, shown in Figure 7-14, combines two serial SCRs sharing a lateral pnp used as the mirror axis of the structure. The MILSCR provides protection for both positive (SCR2) and negative (SCR1) ESD stress polarities. The *npn* bipolar transistor Q1 of SCR1 has a *p*-well diffusion as base, an n^+ diffusion as emitter, and the epi-layer as collector. The shared transistor Q3 is a lateral *pnp* bipolar transistor whose base is the epi-layer and whose emitter and collector are, respectively, Q1 and Q2 base or the reverse, according to the ESD stress polarity. Similarly, Q2, the *npn* bipolar transistor of SCR2, is formed by a *p*-well diffusion as base, an n^+ diffusion as emitter, and the epi-layer as collector. In the PS mode (positive ESD voltage polarity to V_{SS}), the emitter base junction of Q3 is forward biased by V_{ESD}, the ESD voltage, resulting in the bias of Q3 base (epi-layer) to 0.6 V. When the Q3 base voltage reaches the avalanche breakdown voltage of the base-collector junction of Q2, SCR2 triggers on. Once it's triggered the structure provides a very low impedance path to V_{SS}, and the output pad voltage is clamped at the SCR2 holding voltage plus 0.6 V. In the NS mode (negative ESD voltage polarity to V_{SS}), SCR1 is triggered by the avalanche breakdown of the base-collector junction of Q1.

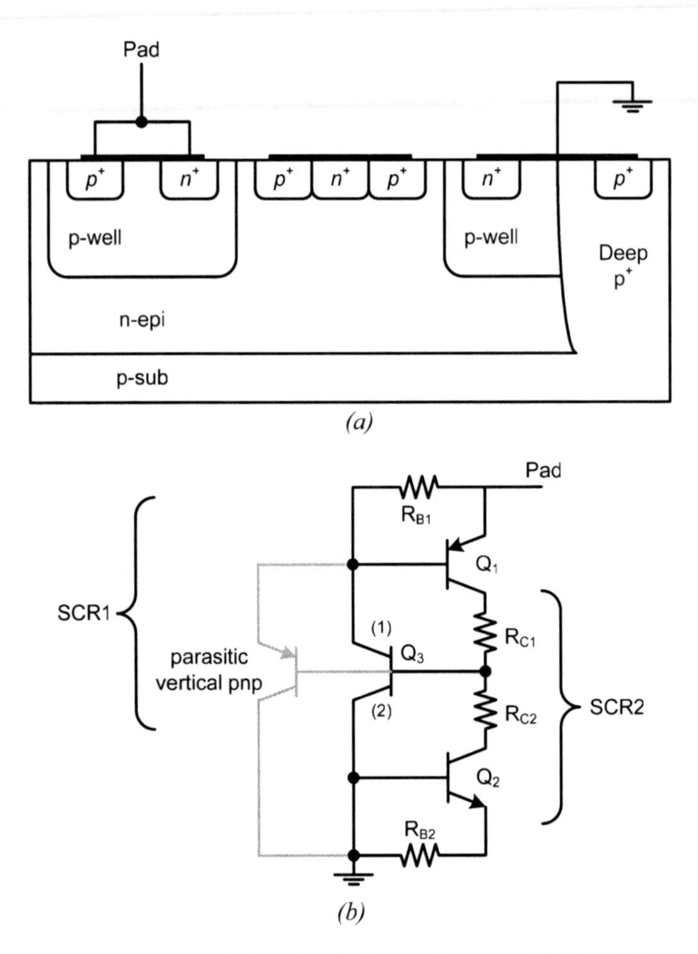

Figure 7-14. (a) Cross-section of MILSCR, (b) Electrical circuit of MILSCR: the emitter of Q_3 is electrode (1) in the PS mode and electrode (2) in the NS mode. (Adapted from [11].)

The MILSCR ESD protection structure was fabricated in a 1.4 μm non-silicided smart-MOS process using the epitaxial substrate [11]. This bi-directional SCR device has a compact structure (75×100 μm^2) with a surface-area reduction of about 15% compared to the conventional dual SCR structure proposed in [16]. The *I-V* characteristics of the MILSCR measured with a curve tracer shown in Figure 7-15. The TLP measurements found that in the PS mode, the triggering voltage is 90 V and the holding voltage is 2.5 V. The device is switched off when the current level decreases below 17 mA, the holding current. In the NS mode, the measured triggering and holding voltages are, respectively, 46 and 1.7 V. The holding current is 5 mA. The failure threshold of MILSCR was determined by step stressing the stand-alone protection device for positive and negative ESD polarities. All samples successfully passed the ESD test up to 10 kV HBM [9, 11].

The main disadvantage of proposed MILSCR structure is relatively low holding voltage and holding current. There is the risk that ESD device can be triggered at normal operating conditions. To increase the holding currents of SCR1 and SCR2 structures in MILSCR device, the base resistances of Q1 and Q2 (Figure 7-14(b)) should be reduced, for example, by implementing a p^+ base contacts [11].

5. POWER BUS ESD PROTECTION CIRCUITS FOR HIGH VOLTAGE APPLICATIONS

In CMOS ICs, the V_{DD}-to-V_{SS} ESD clamp circuits across the power lines of CMOS ICs had been used to effectively increase ESD robustness of the chip. When the ESD protection device is used in the power-rail ESD clamp circuit, the device is expected to be kept off in normal circuit operating condition. During ESD stress conditions, the ESD protection device should be triggered on to discharge ESD current. If the holding voltage of the ESD protection device in power-rail ESD clamp circuit is smaller than the power supply voltage, the ESD device may be triggered on by the system-level electromagnetic compatibility (EMC)/ESD transient pulses to cause latchup failure in CMOS ICs. This phenomenon often leads to IC function failure or even destruction by burning out [18]. Hence, ESD power clamps should have relatively high holding voltage and high holding current to avoid latchup during normal operating conditions.

5.1 ESD Power Clamp-Based on the Field-Oxide Device (FOD)

A latchup-free design of the power-rail ESD clamp circuit with stacked-field-oxide structure was successfully verified in a 0.25 μm 40 V CMOS process [1, 15]. The cross-section of developed high-voltage FOD device structure is shown in Figure 7-15(a). This device is isolated by the n^+ buried layer (NBL) from the common p-type substrate. The base region is inserted between the emitter region and collector region in the layout structure. The spacing between the collector region and the emitter region of the FOD device was 6 μm. The TLP measured I-V curve of developed device is depicted in Figure 7-15(b). This figure shows that the triggering voltage of FOD device is approximately 20 V, the holding voltage is approximately

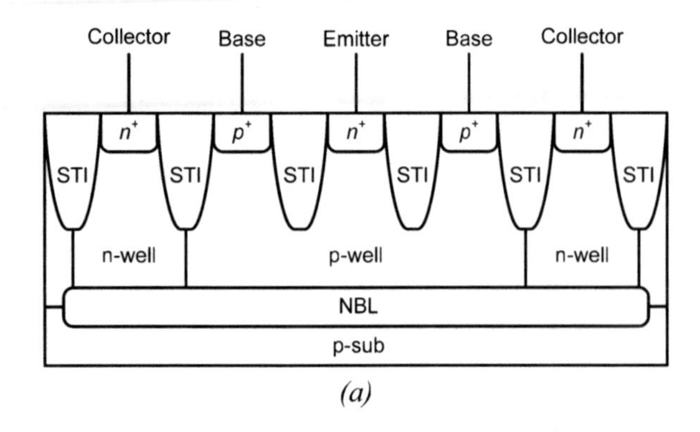

(a)

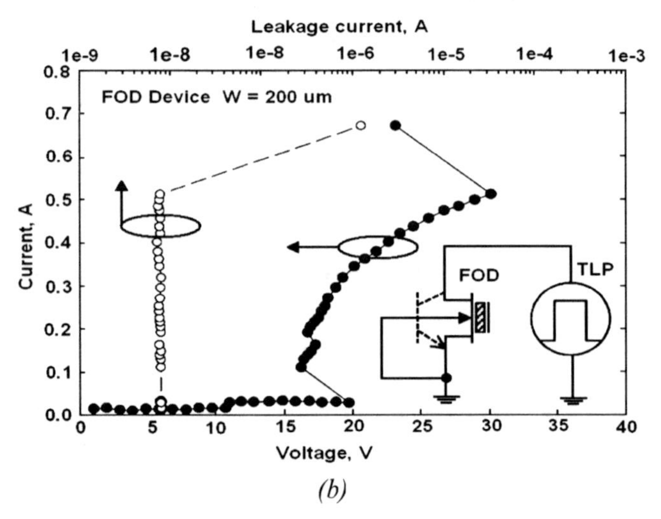

(b)

Figure 7-15. (a) Cross-section of high voltage field-oxide (FOD) device, (b) TLP measured I-V characteristics of FOD device. (Adapted from [15].)

16 V and the second breakdown current of the FOD device with the 200 μm width is 0.5 A. Typically, a transient latchup test (TLU) is used to investigate the susceptibility of the ESD protection devices to the noise transient or glitch on the power lines during normal circuit operating condition. TLU tests shown that the clamped voltage level of the FOD device is approximately 16 V. Hence, the latchup-like issue can be a problem if the single FOD device is used as the power-rail ESD clamp in high-voltage CMOS ICs. Generally, if the holding voltages of high-voltage ESD protection devices are smaller than the power supply voltage under normal circuit operating conditions, the high-voltage CMOS ICs will be susceptible to the latchup danger in the system applications, which often meet the noise or transient glitch issues.

To overcome the latchup related issue between the power rails in high-voltage CMOS ICs during normal circuit operating conditions, a stacked-field-oxide structure has been designed to increase the total holding voltage. It was found that the holding voltage of the stacked-field-oxide structure can be linearly increased by increasing the numbers of cascaded FOD devices [1, 15]. TLP measurements shown that two stacked FOD devices with a device width of approximately 650 μm for each can sustain the typical 2 kV HBM ESD stress and have the holding voltage approximately 35 V. Also, it was found that there is no significant difference on the holding voltage of the stacked-field-oxide structure when the temperature increases from 25°C to 125°C.

Under ESD stress conditions, the ESD clamp device should turn on quickly to bypass the ESD current before the internal circuits are damaged by the ESD energy. Note, that the trigger voltage of the stacked-field-oxide structure is increased in comparison with that of a single FOD device. To reduce the triggering voltage of stacked FOD devices and to reduce its turn-on time, the substrate triggering technique can be used. The proposed latchup-free ESD power clamp with two cascaded FOD devices is shown in Figure 7-16 [15]. Each FOD device in the stacked-field-oxide structure is isolated by the NBL region from the common *p*-type substrate. With two cascaded FOD devices, the total holding voltage of stacked-field-oxide structure in the snapback region is double of that of single FOD device. The latchup immunity of the power-rail ESD clamp circuit to the noise transient during normal circuit operating conditions can be highly increased.

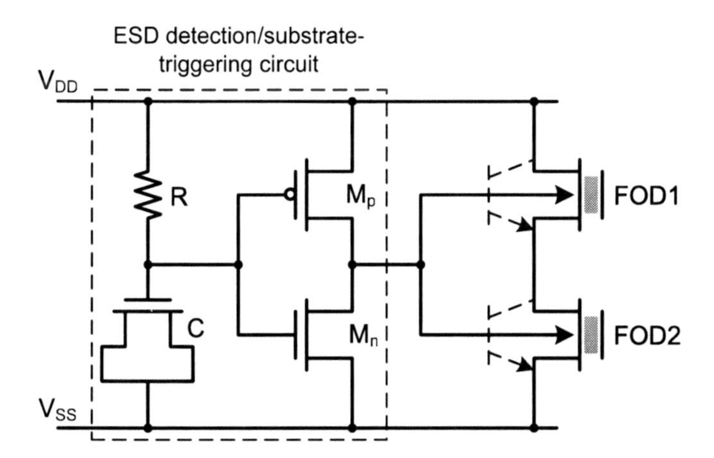

Figure 7-16. Power-rail ESD clamp circuit with two cascaded FOD devices for high-voltage CMOS ICs.

5.2 SCR-Based ESD Power Clamps

The disadvantage of FOD based ESD clamps is that this protection circuits require a relatively big FOD devices. For example, to protect internal devices from 2 kV HBM ESD stress, the width of stacked FOD devices should be 650 μm each. Note, that some high voltage applications require the ESD robustness of 4–8 kV. Hence, FOD based ESD clamps become impractical for such applications. Typically, power clamps with high ESD robustness are based on SCR protection elements. The schematic of the hot-carrier triggered SCR (HCRSCR) ESD protection circuit is shown in Figure 7-17 [19]. Primary ESD protection for positive polarity stress on V_{DD} with respect to V_{SS} is provided by an SCR comprised of bipolar transistors Q1 and Q2 formed using *npn* and *pnp* devices inherent to CMOS technology. The resistances of the *n*-well and *p*-well are represented by R_{NW} and R_{PW} respectively. The SCR is triggered into its low impedance state by hot-carrier generated substrate current injected from NMOS transistor M1, into the base of Q1. Transistors M2–M5 control the triggering of the SCR by turning on the MOSFET M1, and allowing the hot-carrier generation only during an ESD event. Transistor M2 is used as a capacitor to couple the gate of M1, to the positive power supply V_{DD} when it ramps up. The gate of M1 is discharged to V_{SS} by M3 which is turned on by M5. The geometries of M2 and M3 are optimized to ensure that V_{gate} of M1 is greater than its threshold voltage V_{th} during the ESD stress. During normal operation, transistor M3 holds the gate of M1 at V_{SS} and prevents the triggering of SCR. Transistor M4 is used to limit the voltage across the gate oxide of M2.

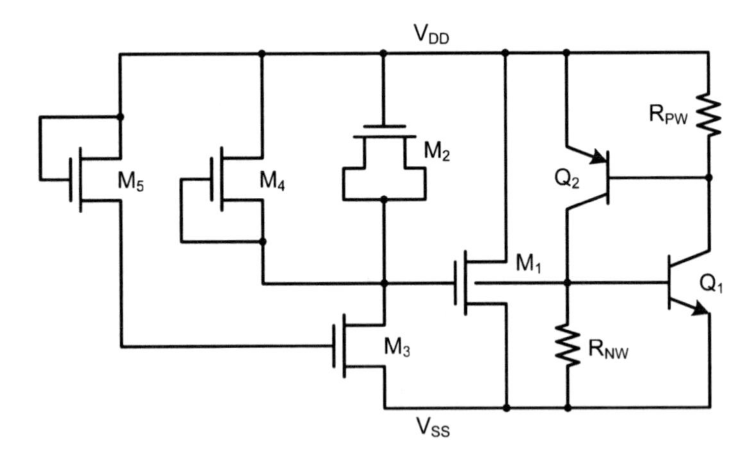

Figure 7-17. Schematic of hot-carriers triggered SCR (HCTSCR) ESD protection circuit.

For negative polarity stress on V_{DD} with respect to V_{SS}, ESD protection is provided by the diode formed by the N-well base of Q2 in the P-type substrate.

The HCTSCR ESD power clamp implemented in 0.5 μm silicided CMOS technology had more than 8 kV of protection level at HBM ESD stress and more than 2 kV of protection level at CDM ESD stress [19]. The width of the SCR and trigger MOSFET M was 88 μm. The trigger voltage of HCTSCR can be tuned by the varying of the gate length of trigger MOSFET M1. The Figure 7-18 shows measurement data of the HCTSCR triggering voltage as a function of the MOSFET M1 gate length L. The data can be fit by the following equation: $V = a + b \cdot log(L)$, where a and b are the process dependent constants. This equation allows determining the optimal gate length of M1 to ensure an adequate level of ESD protection and preventing the triggering of the SCR during normal circuit operation.

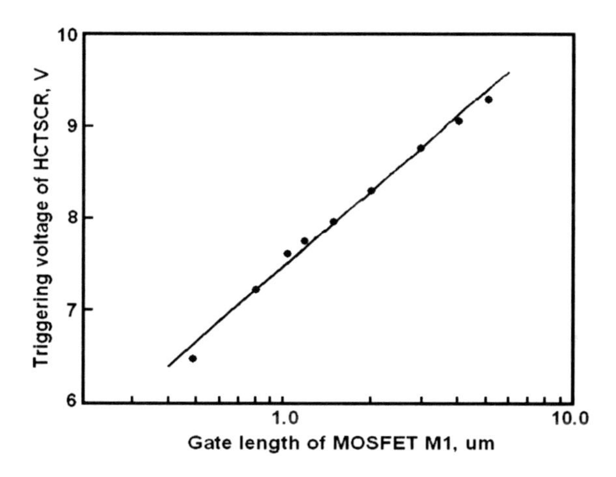

Figure 7-18. Triggering voltage of HCTSCR vs. Gate length of MOSFET M_1. (Adapted from [19].)

For smart power applications, which require the high level of failure threshold of ESD power clamps, the embedded SCR LDMOS (ESCR-LDMOS) structure was developed [20]. At normal operating conditions, the DC *I-V* characteristics of this device are typical for LDMOS transistor. However during the ESD stress, the SCR of ESCR-LDMOS is turned on and provides a low impedance path to discharge the ESD energy. The cross-section and equivalent electrical circuit of the ESCR-LDMOS structure is shown in Figure 7-19. Typically, the ESD MOS devices are often designed with large width drain for creating a ballast resistor to have uniform current

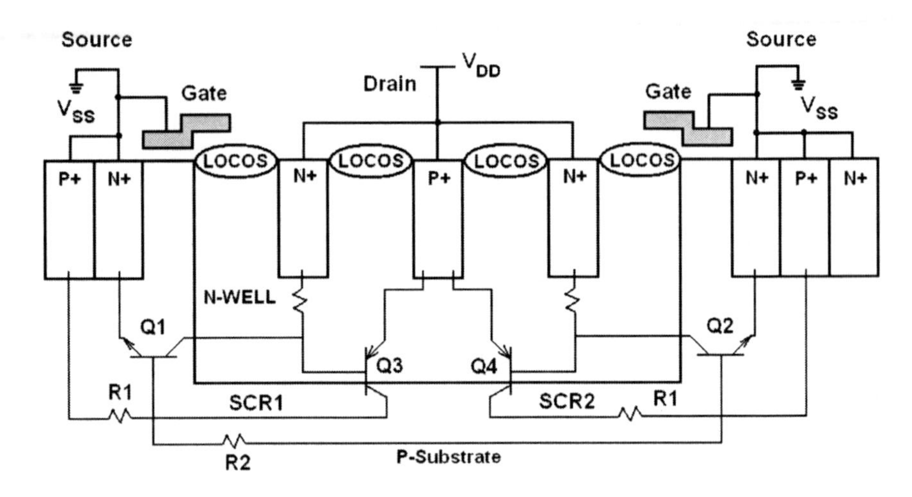

Figure 7-19. Cross-section and equivalent electrical circuit of the ESCR-LDMOS structure with a p^+ strap in source region.

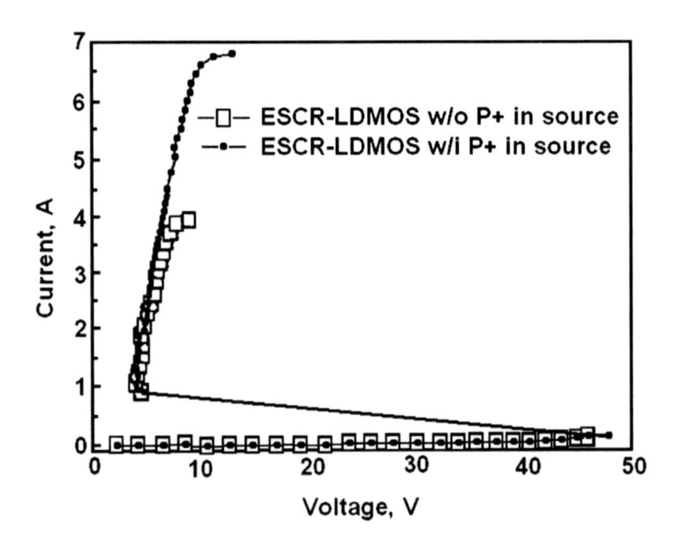

Figure 7-20. TLP measured I-V characteristics of the ESCR-LDMOS ESD clamp with and without p^+ straps in source region.

distribution in the multi-fingers device. The ESCR-LDMOS has a wide enough drain region to sacrifice part of n^+ strap by inserting the p^+ strap into the drain region and without increasing the device dimension.

The failure analysis of ESCR-LDMOS structure after HBM ESD stress shown the non-uniform current distribution in multi-fingers ESCR-LDMOS due to the effective P-substrate shunting resistance varying with the finger position. To improve the current uniformity, the P^+ straps were inserted into each source region of ESCR-LDMOS, as depicted in Figure 7-19. The TLP measured *I-V* characteristics of the ESCR-LDMOS structure is presented in Figure 7-20. A 200 μm width ESD clamp was implemented in 40 V CMOS process. Its ESD failure thresholds at HBM ESD stress and MM stress were more than 8 kV and 800 V, respectively [20].

6. SUMMARY

Smart power technology allows the integration of power devices and both analog and digital circuitry on the same chip. Such device integration makes possible the implementation of fully integrated systems (for example, car injection control system) with a significant increase in reliability and packaging capabilities. One of the typical applications of smart power technology is the automotive electronics. Automotive companies require a very high quality ICs (a significant shift from PPM (parts per million) defect level to 'Zero Defects') with a high performance and significant cost savings. As today's microelectronics technology continues shrink to sub-micron/deep-sub-micron dimensions, ESD damage in integrated circuits has become one of major reliability issues, especially in the harsh automotive environment. This chapter described the typical devices and circuits, which are used for ESD protection in smart power ICs. Since smart power technology uses MOS and BJT transistors on the same chip, the ESD protection strategy can be based on the LDMOS, BJT or SCR protection devices. Typically, SCR based ESD protection circuits have a high ESD robustness. However, the latch-up issues can be a serious obstacle for using these protection circuits for smart power applications. By this reason, a special power bus ESD protection circuits optimized for high voltage applications were also considered in this chapter.

REFERENCES

[1] K.-H. Lin and M. D. Ker, "Design on latchup-free power-rail ESD clamp circuit in high-voltage CMOS ICs," *EOS/ESD Symposium*, pp. 265–272, 2004.

[2] C. Duvvury, F. Carvajal, C. Jones, and D. Briggs, "Lateral DMOS design for ESD robustness," *Proc. of the Int. Electron Devices Meeting (IEDM)*, pp. 375–378, 1997.

[3] G. Meneghesso, M. Ciappa, P. Malberti, L. Sponton, G. Croce, C. Contiero, and E. Zanoni, "Overstress and electrostatic discharge in CMOS and BCD integrated circuits," *Microelectronics Reliability*, vol. 40, No. 8–10, pp. 1739–1746, 2000.

[4] M. P. J. Mergens, W. Wilkening, S. Mettle, H. Wolf, A. Stricker, and W. Fichtner, "Analysis of lateral DMOS power devices under ESD stress conditions," *IEEE Trans. on Electron Devices*, vol. 47, No. 11, pp. 2128–2137, 2000.

[5] L. Sponton, L. Cerati, G. Croce, F. Chrappan, C. Contiero, G. Meneghesso, and E. Zanoni, "ESD protection structures for BCD5 smart power technologies," *Microelectronics Reliability*, vol. 41, No. 9–10, pp. 1683–1687, 2001.

[6] T. Polgreen, and A. Chatterjee, "Improving the ESD failure threshold of silicided n-MOS output transistors by ensuring uniform current flow," *IEEE Trans. on Electron. Devices*, vol. 39, No. 2, pp. 379–388, 1992.

[7] F. Marchio, V. Poletto, A. Russo, G. Torrisi, J. Notaro, G. Buriak, and M. Mirowski, "A R-evolution in power electronics: From "Intelligent power" to "Super smart power" in automotive," *Proc. of Power Electronics in Transportation*, pp. 27–34, 2004.

[8] N. Speciale, A. Leone, S. Graffi, and G. Masetti, "A unified approach for modeling multi-terminal bipolar and MOS devices in smart-power technologies," *Proc. of European Solid-State Devices Research Conf. (ESSDERC)*, pp. 312–315, 1997.

[9] C. Delage, N. Nolhier, M. Bafleur, J. M. Dorkel, J. Hamid, P. Givelin, and J. Lin-Kwang, "The mirrored lateral SCR (MILSCR) as an ESD protection structure for smart power applications," *Proc. of IEEE Bipolar/BiCMOS Circuits and Technology Meeting (BCTM)*, pp. 191–194, 1998.

[10] G. Bertrand, C. Delage, M. Bafleur, N. Nolhier, J. -M. Dorkel, Q. Nguyen, N. Mauran, D. Tremouilles, and P. Perdu, "Analysis and compact modeling of a vertical grounded-based n-p-n bipolar transistor used as ESD protection in a smart power technology," *IEEE J. of Solid-State Cir.*, vol. 36, No. 9, pp. 1373–1381, 2001.

[11] C. Delage, N. Nolhier, M. Bafleur, J. -M. Dorkel, J. Hamid, P. Givelin, and J. Lin-Kwang, "The mirrored lateral SCR (MILSCR) as an ESD protection structure: Design and optimization using 2-D device simulation," *IEEE Trans. of Solid-State Cir.*, vol. 34, No. 9, pp. 1283–1289, 1999.

[12] V. De Heyn, G. Groeseneken, B. Keppens, M. Natarajan, L. Vacaresse, and G. Gallopyn, "Design and analysis of new protection structures for smart power technology with controlled trigger and holding voltage," *Proc. of the Int. Reliability Physics Symp. (IRPS)*, pp. 253–258, 2001.

[13] V. A. Vashchenko, A. Concannon, M. ter Beek, and P. Hopper, "High performance SCRs for on-chip ESD protection in high voltage BCD processes," *Proc. of the IEEE Power Semiconductor Devices and ICs (ISPSD)*, pp. 261–264, 2003.

[14] V. A. Vashchenko, A. Concannon, M. ter Beek, and P. Hopper, "Comparison of ESD protection capability of lateral BJT, SCR and bidirectional. SCR for hi-voltage BiCMOS circuits," *Proc. of the IEEE Bipolar/BiCMOS Circuits and Technology Meeting (BCTM)*, pp. 181–184, 2002.

[15] M. D. Ker and K. H. Lin, "The impact of low-holding-voltage issue in high-voltage CMOS technology and the design of latchup-free power-rail ESD clamp circuit for LCD driver ICs," *IEEE J. of Solid-State Cir.*, vol. 40, No. 8, pp. 1751–1759, 2005.

[16] C. -Y. Wu, M. D. Ker, C. -Y. Lee, and L. Ko, "A new on-chip ESD protection circuit with dual parasitic SCR structures for CMOS VLSI," *IEEE J. of Solid-State Cir.*, vol. 27, No. 3, pp. 274–280, 1992.

[17] C. -Y. Huang, W. -F. Chen, S. -Y. Chuan, F. -C. Chiu, J.C. Tseng, I. -C. Lin, C. -J. Chao, L. -Y. Leu, and M. D. Ker, "Design optimization of ESD protection and latchup prevention for a serial I/O IC," *Microelectronics Reliability*, vol. 44, No. 2, pp. 213–221, 2004.

[18] R. Lewis and J. Minor, "Simulation of a system level transient-induced latchup event," *EOS/ESD Symposium*, pp. 193–199, 1994.

[19] J. T. Watt and A. J. Walker, "A hot-carrier triggered SCR for smart power bus ESD protection," *Proc. of the Int. Electron Devices Meeting (IEDM)*, pp. 341–344, 1995.

[20] J. -H. Lee, J. R. Shih, C. S. Tang, K. C. Liu, Y. H. Wu, R. Y. Shiue, T. C. Ong, Y. K. Peng, and J. T. Yue, "Novel ESD protection structure with embedded SCR LDMOS for smart power technology," *Proc. of the Int. Reliability Physics Symp. (IRPS)*, pp. 156–161, 2002.

Chapter 8

ESD PROTECTION FOR RF CIRCUITS

1. INTRODUCTION

Wireless communication market is growing rapidly. In the first half of the twentieth century, wireless communication in the form of radio and television had the most significant impact on everybody's life. In the second half of the century, the rapid development of semiconductor devices and integrated circuits lead to a much wider applications of wireless communication systems. There is an increasing demand for portable electronic devices such as cellular and cordless phones, pagers, wireless modems and GPS receivers. Furthermore, more options are constantly added to these portable devices. As a result, today's cellular phones have a much wider capabilities such as sending and receiving data, pictures and even receiving radio and television networks. This evolution is a result of constant increase in the integration level of semiconductor devices and reduction in their cost.

There are various semiconductor technologies available for radio frequency applications. GaAs and silicon CMOS, BiCMOS and bipolar technologies are the most widely used technologies nowadays. In traditional RF circuits, GaAs, bipolar and ceramic SAW filters were used for the RF section, bipolar for the IF section and CMOS for base band. However, as CMOS technology scales into deep submicron regime, higher operating frequency of MOS transistors allows the design of a fully integrated CMOS System on Chip (SoC).

O. Semenov et al., ESD Protection Device and Circuit Design for Advanced CMOS Technologies, 199–218.

In all of the above mentioned systems, the receiver block is one of the most challenging parts of the design. There are many different receiver architectures in the literature. The most common receiver architectures are based on superheterodyne receiver, which was invented in 1918 by Armstrong [1]. This method is based on down converting the input frequency to an intermediate frequency which is usually called IF frequency. Figure 8-1 shows the block diagram of a superheterodyne receiver.

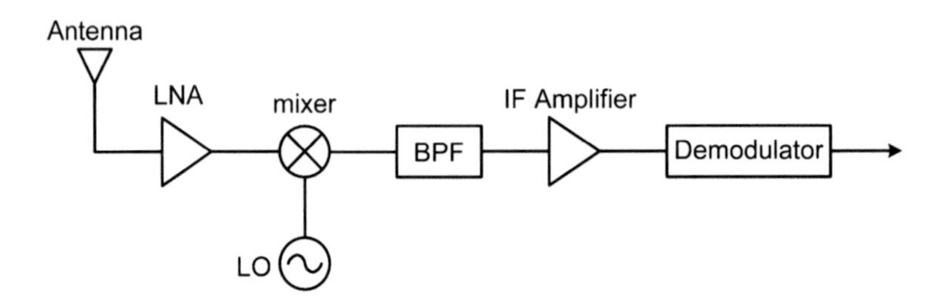

Figure 8-1. Superheterodyne receiver.

In this block diagram, the input RF signal is received at the antenna and is amplified by the Low Noise Amplifier (LNA). Down conversion of RF frequency is achieved using the Local Oscillator (LO) and the mixer. The output frequency of the mixer is called the Intermediate Frequency (IF) and is amplified with the IF amplifier. Finally, output of the IF amplifier is sent to the demodulator.

As any integrated circuit, RF systems need to have an ESD protection circuit. Referring back to Figure 8-1 it can be seen that the LNA is the first block of the system. This block is the most sensitive one to parasitics of ESD protection circuits. An LNA block has input matching and low noise requirements. However, adding an ESD protection circuit to the LNA adds extra parasitic capacitance and resistance to the input of the amplifier. These extra capacitance and resistance degrade both matching constraint and noise requirements of the LNA. Therefore, similar to the discussions in Chapter 6, the parasitics of ESD protection circuit should be minimized.

Although reducing parasitic capacitance of the ESD protection circuit is a general requirement to design ESD protection for high speed applications, there are other methods specific to narrow-band LNAs. This chapter is dedicated to ESD protection strategies for low noise amplifiers. This chapter is organized as follows: In Section 2 basic definitions that are used in RF applications are defined. Low noise amplifier architectures are discussed in Section 3. Finally, ESD protection techniques for LNAs are presented in Section 4.

2. BASIC CONCEPTS

In this section, before going through the details of ESD protection for RF circuits, a few basic concepts in RF circuits, especially low noise amplifiers (LNAs), are discussed. These concepts include quality factor of inductors and capacitors, Signal to Noise Ratio (SNR), Noise Figure (NF), matching and s-parameters and gain of amplifiers.

2.1 Quality Factor of Inductors and Capacitors

Inductors and capacitors in integrated circuits have non-idealities mainly in the form of ohmic resistance. As a result, unlike a purely reactive element, the phase difference between voltage and current is not precisely 90°. This ohmic resistance increases the power dissipation and thermal noise of the reactive element. Quality factor is a parameter that defines the purity of the reactive element and is defined as the ratio between average reactive power and average dissipated power. If a real reactive element is defined as an ideal reactive element in series with a resistor R_S, quality factor Q of the reactive element becomes:

$$Q = \frac{X_S}{R_S} \tag{8-1}$$

where X_S is the reactance of the capacitor or inductor at the given frequency. Hence, quality factor of inductor and capacitor becomes:

$$Q_L = \frac{\omega L}{R_S} \quad \text{and} \quad Q_C = \frac{1}{\omega C R_S} \tag{8-2}$$

Now, consider a series resonant RLC network. This network is characterized by its resonant frequency and the quality factor. The resonant frequency ω_0 and quality factor Q at this frequency for this network are calculated from the following equations [2]:

$$\omega_0 = \frac{1}{\sqrt{LC}} \tag{8-3}$$

$$Q = \frac{\omega_0 L}{R} = \frac{1}{\omega_0 CR} \tag{8-4}$$

2.2 SNR and Noise Figure

The input of the receiver shown in Figure 8-1 receives both the main signal and noise. Both noise and signal are present at different points of the receiver. In order to measure purity of the signal at each point, a parameter called signal to noise ratio (SNR) is defined as follows:

$$SNR = \frac{S}{N} = \frac{\text{signal power}}{\text{noise power}} \tag{8-5}$$

An ideal amplifier amplifies both signal and noise and hence, doesn't change the SNR. However, a real amplifier adds extra noise to the signal which is coming from resistive and active elements of the amplifier. This extra noise causes a reduction in SNR. In an RF receiver, the change in SNR is measured through noise factor (F), which is defined as the ratio between SNR at input to SNR at output:

$$F = \frac{SNR_i}{SNR_o} = \frac{S_i/N_i}{S_o/N_o} \tag{8-6}$$

As mentioned earlier, every block reduces the SNR of its input. Therefore, F is always between 1 and ∞. A more common term to measure noise performance in RF applications is noise figure (NF) which is simply noise factor expressed in dB:

$$NF = 10\ \log(F) = 10\ \log\left(\frac{SNR_i}{SNR_o}\right) \tag{8-7}$$

As a result, noise figure is always greater than zero.

2.3 Impedance Matching

Another concept that is very important in RF circuits is the impedance matching. Impedance matching networks are designed with two main objecttives: matching for noise and matching for power. In the former, the main objective is to design the matching network to obtain minimum noise figure, while in the latter, the purpose of the matching network is to maximize signal power. Both of these two designs have advantages and disadvantages. However, the latter is becoming more dominant in today's systems. The reason is that different blocks of an RF system are designed with standard termination impedance to be able to be used in various systems. The standard termination impedance for RF systems is 50 Ω.

Based on the above discussion, matching networks are designed based on the maximum power transfer theorem [2]. Consider a load impedance Z_L is connected to a source with series impedance Z_S. Z_L and Z_S are defined in equation (8-8).

$$Z_L = R_L + jX_L, \; Z_S = R_S + jX_S \tag{8-8}$$

This theorem states that, in order to maximize the power delivered to the load, load impedance Z_L should be complex conjugate of Z_S, i.e. $R_L = R_S$ and $X_L = -X_S$.

In RF systems, impedance matching is provided with passive elements. L-matching network is the simplest form of matching network, which is shown in Figure 8-2.

This matching network can be used to either increase or decrease the impedance. Figure 8-2(a) shows an upward impedance transformer. This block increases the resistance R_S and converts it to Z. Similarly, by changing the connection of the inductor and capacitor, downward impedance transformer is implemented as shown in Figure 8-2(b).

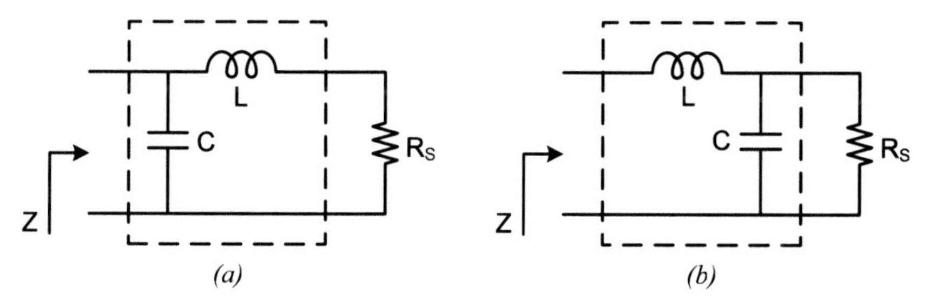

Figure 8-2. L-matching network (a) upward impedance transformer (b) downward impedance transformer.

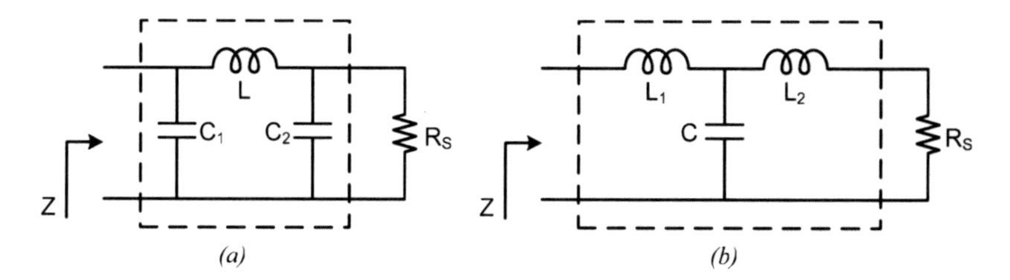

Figure 8-3. (a) π-type impedance transformer. (b) T-type impedance transformer.

To increase the degree of freedom in designing the matching network, π-type and T-type matching networks can be used. Figure 8-3 shows the schematic of these matching networks. Both of these networks can be considered as two L-type matching networks connected together. In the π-type network, R_S is first increased by C_2-L network and then decreased by C_1-L network. Similarly, T-type network first decreases R_S with C-L_2 network and then increases it with C-L_1 network. The benefit of the π-type is that parasitic capacitance of input and output interfaces can be designed as part of the matching network. On the other hand, T-type network is more useful when the termination parasitics are inductive instead of capacitive.

2.4 S-Parameters

Behavior of a two port network can be modeled in many different ways. At low frequencies, impedance (z), admittance (y) and hybrid (h) are examples of the most common models [2]. In these models, all parameter measurements are based on creating short-circuit and open-circuit configurations at both input and output ports. However, at higher frequencies achieving short-circuit and open-circuit over a broadband range becomes very difficult. Moreover, active high frequency circuits may even oscillate when terminated in open or short circuit configurations. Hence, in order to deal with these experimental problems, in high frequencies, scattering parameters (s parameters) are used to characterize a two port circuit. This model is based on measuring the incident and reflected power at all ports of a network. Let's consider a two port block as shown in Figure 8-4.

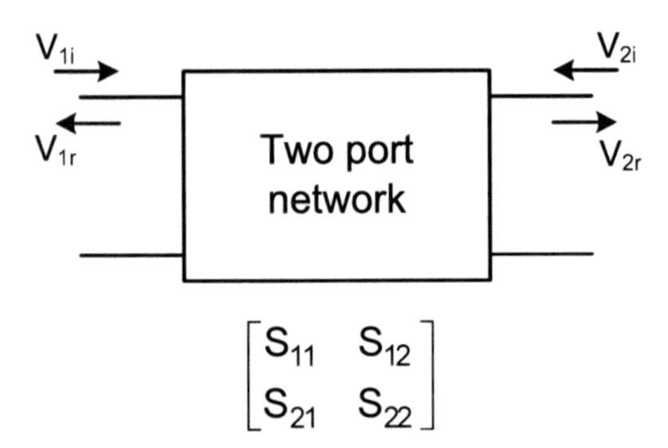

Figure 8-4. s-parameters of a two-port system.

In Figure 8-4 V_{1i} and V_{2i} are the incident voltages to two ports of the network while V_{1r} and V_{2r} are the reflected voltages from the two ports. The values of the s-parameters are defined as follows:

$$S_{11} = \frac{\text{reflected power at port 1}}{\text{incident power at port 1}}; \text{ when incident power at port 2 is 0} \quad (8\text{-}9)$$

$$S_{12} = \frac{\text{reflected power at port 1}}{\text{incident power at port 2}}; \text{ when incident power at port 1 is 0} \quad (8\text{-}10)$$

$$S_{21} = \frac{\text{reflected power at port 2}}{\text{incident power at port 1}}; \text{ when incident power at port 2 is 0} \quad (8\text{-}11)$$

$$S_{22} = \frac{\text{reflected power at port 2}}{\text{incident power at port 2}}; \text{ when incident power at port 1 is 0} \quad (8\text{-}12)$$

In definition of s-parameters, making the incident power zero at one port is equivalent to creating match condition on that port. For example, to measure S_{11}, port 2 is terminated with a load impedance equal to characteristic impedance of the port. As a result, load at port 2 absorbs all the power and reflected power of the load becomes zero. Therefore, incident power at port 2 becomes zero as well and the ratio of the reflected power to incident power at port 1 represents S_{11}.

In order to have a better understanding of the s-parameters, let's assume that the two port network shown in Figure 8-4 is an amplifier. In this network, S_{11} and S_{22} represent the input and output impedance of the amplifier, S_{21} represents the forward gain and finally, S_{12} represents the reverse transmission gain. The value of S_{12} of an amplifier is usually very small at lower frequencies but it becomes significant at higher frequencies.

3. LOW NOISE AMPLIFER

Low Noise Amplifier (LNA) is the first stage in most receiver architectures, as shown in Figure 8-1. This LNA should meet certain requirements to be useful for an RF system. The most important requirements for an LNA are matching, noise and gain.

Matching is one of the most important requirements of an LNA. As LNA is fed through a 50 Ω transmission line coming from the antenna, input impedance of the LNA should be 50 Ω to avoid reflections over the transmission line. Noise is another important consideration in the design of an LNA. As LNA is the first stage after the antenna it sets the lower bound on the achievable noise figure of the entire receiver. Hence, design of an LNA with a very low noise figure is crucial. Finally, as in any amplifier, an LNA should have high voltage gain. Moreover, this high gain is desirable as it improves the SNR of the stages following the LNA.

There are a number of suitable implementations for an LNA. Most common LNA architectures are based on common source and common gate amplifiers. The following two subsections discuss and compare these two topologies.

3.1 Common Source LNA

Common source amplifier is a popular configuration in RF applications. However, as regular common source amplifier has capacitive input impedance, inductors are added to achieve 50 Ω resistive impedance. Figure 8-5 shows the input stage of a common source low noise amplifier.

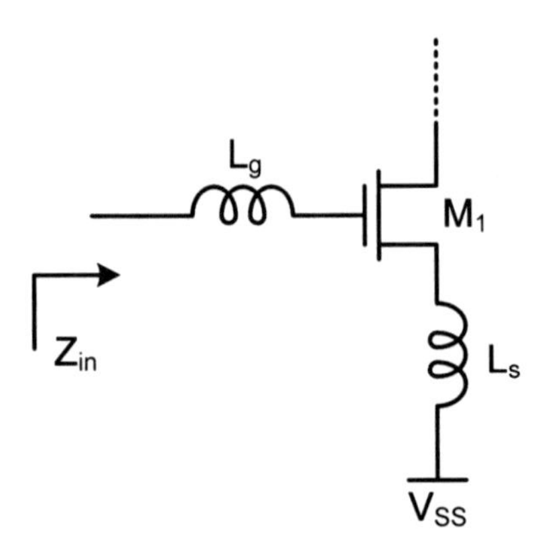

Figure 8-5. Input stage of a common source LNA.

Using small signal model for transistor M_1 the input impedance Z_{in} of the LNA becomes:

$$Z_{in} = s(L_s + L_g) + \frac{1}{sC_{gs1}} + \left(\frac{g_{m1}}{C_{gs1}}\right)L_s \tag{8-13}$$

where C_{gs1} and g_{m1} are the gate-source capacitance and transconductance of the transistor M_1.

As mentioned earlier, this impedance should be purely resistive and equal to 50 Ω for the given frequency band of interest. The impedance Z_{in} becomes purely resistive at resonance. Hence, the resonance frequency and input impedance at resonance frequency for this LNA are calculated from the following equations:

$$\omega_r = \frac{1}{\sqrt{C_{gs1}\left(L_s + L_g\right)}} \tag{8-14}$$

$$Z_{in}\left(\omega = \omega_r\right) = \left(\frac{g_{m1}}{C_{gs1}}\right)L_s \tag{8-15}$$

Figure 8-6 shows the schematic of the most popular low noise amplifier which is called inductively degenerated common source LNA. The input stage is the same as Figure 8-5 and hence, resonance frequency and input impedance are calculated from equations (8-14) and (8-15).

The performance of a low noise amplifier is measured with s-parameters. Design target for an LNA is to achieve 50 Ω input and output impedance with maximum gain and best noise performance. Input and output impedance of the LNA are measured with S_{11} and S_{22} respectively. Power gain of the amplifier is measured with S_{21}, while noise figure represents noise performance of the amplifier. Table 8-1 shows typical performance of state of the art low noise amplifiers.

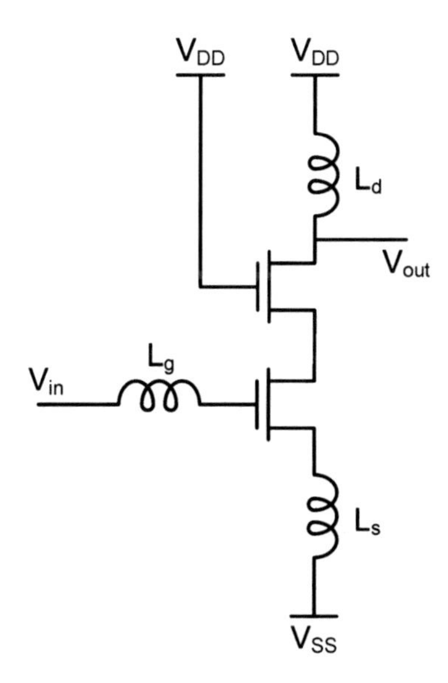

Figure 8-6. Inductively degenerated common source LNA.

Table 8-1. Typical common source LNAs.

	[3]	[3]	[4]
Technology	0.35 μm	0.35 μm	0.18 μm
Frequency	0.92 GHz	0.92 GHz	5.75 GHz
Input matching (S_{11})	–10.1 dB	–19 dB	–21 dB
Output matching (S_{22})	–27 dB	–27 dB	–11 dB
Power gain (S_{21})	14 dB	15 dB	14.1 dB
Noise figure (NF)	0.9 dB	1.15 dB	1.8 dB

A simple guideline for the required specifications of a low noise amplifier can be summarized as follows: S_{11} and S_{22} should be less than –10 dB, NF should be less than 3 dB and power gain must be as high as possible.

3.2 Common Gate LNA

Common gate LNA is the main competitor of the inductively degenerated common source LNA. Figure 8-7 shows the schematic of the common gate LNA. In this schematic input impedance consists of the inductor L_s in parallel with the gate-source capacitance of M_1 (C_{gs1}) in parallel with inverse transconductance of M_1 ($1/g_{m1}$). As a result, in resonance, a good approximate of the input impedance is simply $1/g_{m1}$. As the input impedance of the LNA should be 50 Ω, transistor M_1 can be designed to satisfy this requirement.

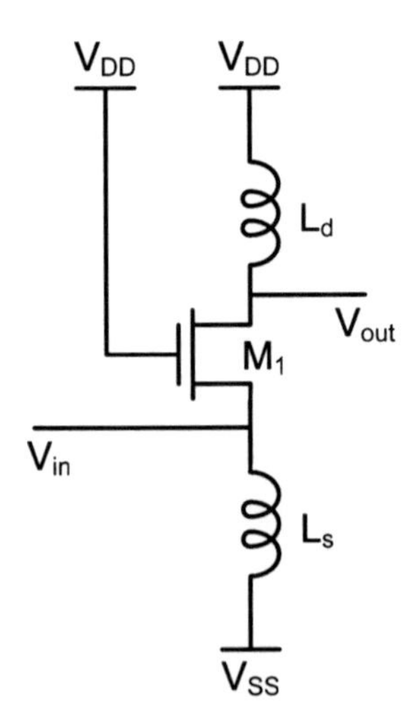

Figure 8-7. Common gate LNA.

It has been shown that one of the main disadvantages of the common gate LNA is that its input impedance depends on drain source resistance (r_{ds}) of M_1 [2].

4. ESD PROTECTION METHODS FOR RF CIRCUITS

In an LNA, input pins are the most susceptible pins to ESD which is due to their connection to gate of MOS transistors. At the same time, adding an ESD protection circuit to input and output pads of the LNA adds parasitic capacitance and resistance to input and output pads of the LNA. These extra parasitics degrade input/output matching, gain and noise of the LNA significantly. The high sensitivity of LNA performance to parasitics of ESD protection circuit is creating a bottleneck for introducing CMOS RF circuits to the market.

Before going through the details of ESD protection methods in RF applications, let's consider the problem from a different angel. An LNA needs to face two different signals: RF signal and ESD stress. RF signal should pass through the amplifier smoothly but the ESD stress energy must be kept away from the amplifier. These two signals have main differences that can be used to design a proper ESD protection circuit. The first difference is in their energy or voltage level. ESD signal has a very high voltage level which makes it easy to detect. Snapback-based ESD protection circuits discussed in Chapter 3 use a similar concept. The second difference between RF signal and ESD signal is in their frequency. ESD signal has a relatively low frequency which is in a few tens of MHz range. On the other hand, RF signal has a much higher frequency which is in GHz range. Hence, it should be possible to provide ESD protection by designing a proper passive filter. In the rest of this section some of these protection methods are presented.

4.1 Cancellation Technique

As discussed in previous chapters, snapback-based ESD protection circuits are capable of providing a low parasitics protection. In other words, the ESD protected input pad has extra parasitic capacitance. This additional capacitance can be compensated by adding an inductor in parallel with the protecttion circuit as shown in Figure 8-8 [5].

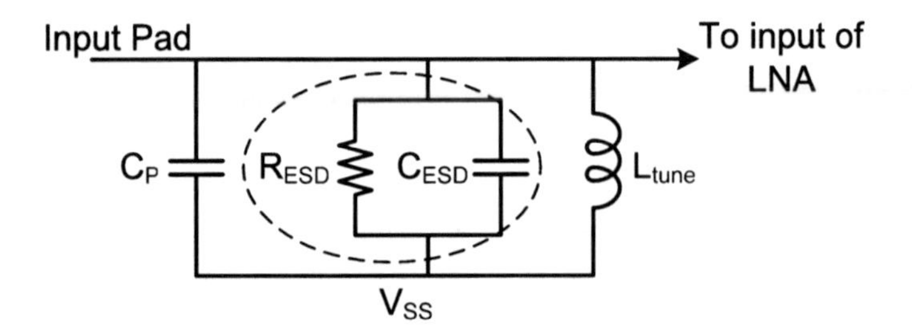

Figure 8-8. Cancellation technique.

In this figure R_{ESD} and C_{ESD} represent the ESD protection circuit, C_P models bond pad capacitance and L_{tune} is the added inductor. Under ESD conditions, the ESD protection circuit, which is shown as C_{ESD} and R_{ESD}, provides the required protection. Under normal operating conditions, L_{tune} creates a resonant circuit with C_P and C_{ESD} to tune out the impact of all the parasitic capacitances. In other words, the value of L_{tune} can be calculated from the following equation:

$$L_{tune} = \frac{1}{(C_P + C_{ESD})\omega_0^2} \tag{8-16}$$

where ω_0 is the operating frequency of the LNA.

This technique was applied to a 5.25 GHz LNA in 0.35 μm BiCMOS technology [5]. ESD protection was provided using grounded gate NMOS, where 3.6 kV HBM protection level was achieved. At the same time, impact on performance of the LNA was minimal, where power gain is un-affected and noise figure is increased from 2.2 dB to only 2.4 dB.

4.2 ESD Protection with a Π-Type Matching Network

ESD protection circuits are mainly designed as separate blocks that are added to pins. However, for RF applications, an ESD-RF co-design approach has been developed [6]. In this approach, parasitic capacitance of the protection circuit is considered as part of the matching network. Therefore, the LNA and ESD protection circuit should be designed and optimized together.

In order to include the ESD protection circuit in input and output matching networks of an LNA, Π-type matching network is used. Figure 8-9 shows the input stage of a common source LNA with a Π-type matching network.

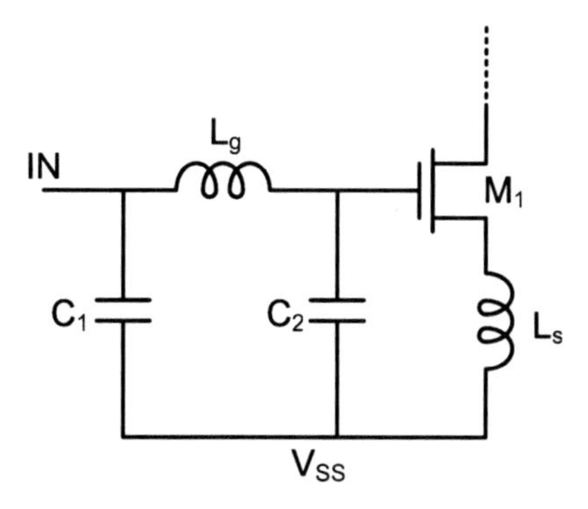

Figure 8-9. Input stage of a common source LNA with Π-type matching network.

It can be seen that compared to a regular common-source LNA, inductor L_g is replaced by a Π-type network consists of L_g, C_1 and C_2. C_2 represents capacitance of the ESD protection circuit, while C1 includes the bond pad capacitance. Moreover, C_1 can be considered as another ESD protection element as well. As mentioned in Section 2.3, one of the features of the Π-type matching network is having an extra degree of freedom, which allows setup the resonant (operating) frequency, bandwidth and impedance-transforming ratio. On the other hand, this method has some limitations. As design of the LNA and ESD protection should be done at the same time, RF model of all components including ESD protection circuits are required. However, in most cases these models are not available. Alternatively, simplistic models with limited accuracy are available. The easiest solution is to extract the small signal model of ESD protection circuits [7] and use it to fine tune performance of the LNA. Normally, several iterations are required to design an LNA with minimum noise figure and maximum gain, while ensuring adequate ESD protection.

The common source LNA with Π-type matching network topology was used to design a 1.9 GHz LNA with 50 Ω input and output impedance match [6]. Figure 8-10 shows the schematic of this LNA.

A buffer is added to LNA to provide better isolation between input and output and hence, makes the design of the output matching easier. ESD protection for this LNA is provided using grounded-gate MOSFET (GGMOS) in the Π-type network. These GGMOS transistors were optimized based on only their ESD performance by designing proper gate length, drain contact to gate spacing and silicide block option. For DC power supplies, ESD protection is also provided using a simple grounded-gate MOSFET which, for simplicity, is not shown in Figure 8-10.

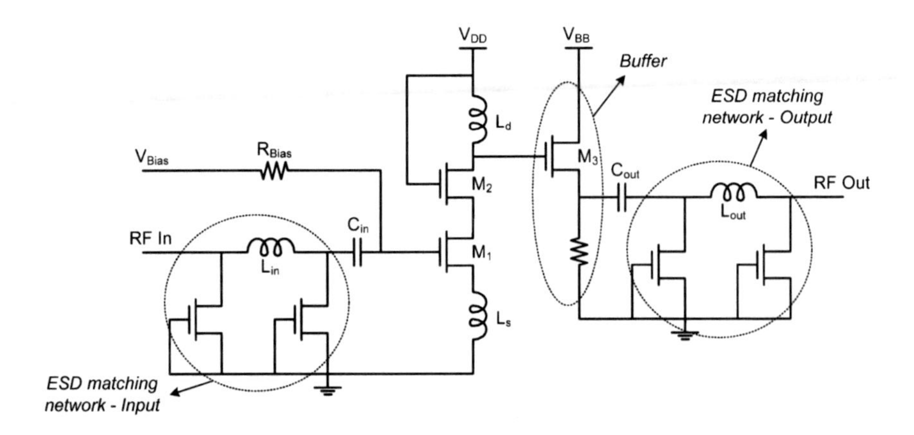

Figure 8-10. LNA with Π-type matched ESD protection.

Table 8-2. Simulated performance of the LNA.

	LNA without ESD	LNA + attached ESD	RF-ESD co-design
Gain, S_{21} (dB)	25.73	19.67	19.62
Input match S_{11} (dB)	–13.93	–10.34	–12.84
Output match S_{22} (dB)	–29.63	–11.5	–16.1
Noise figure (dB)	2.56	4.95	3.85
IIP3 (dBm)	–24.31	–19.45	–20.15
Power (mW)	68	68	68

In order to examine the advantage of the RF-ESD co-design method, the 1.9 GHz LNA is simulated in three different configurations: without ESD protection, with a simple GGMOS protection connected to all pads, and with RF-ESD co-design method as shown in Figure 8-10. Table 8-2 compares simulation results for these three configurations [6]. The first column is the LNA without any ESD protection. The second column is the LNA with attached GGMOS protection and the last column is the RF-ESD co-design method using the same GGMOS protection. It can be seen that adding ESD protection degrades RF performance of the LNA. However, comparing the last two columns of Table 8-2 shows that RF-ESD co-design method improves noise figure and input/output matching of the LNA significantly.

The 1.9 GHz LNA shown in Figure 8-10 was implemented in 0.25 μm CMOS process. ESD robustness of the LNA was tested using HBM measurement. The test was done for all four zapping modes and the LNA passed 3 kV HBM stress. The measured performance of the LNA at 1.9 GHz is shown in Table 8-3. It can be seen that although noise figure and gain of the LNA are worse than the simulated results, good input and output matching are achieved for the amplifier.

Table 8-3. Measured RF performance of the LNA.

S_{21} (dB)	S_{11} (dB)	S_{22} (dB)	NF (dB)
8.7	–14.5	–15	5

4.3 Inductive ESD Protection

So far in this chapter, ESD protection for RF circuits was designed using conventional protection devices, i.e. MOS-based or SCR-based devices. The major constraint was to reduce the impact of parasitic capacitance of the protection circuit on RF circuit performance. In other words, the separation of RF and ESD signals is based on magnitude of the applied voltage. However, as mentioned in Section 4, RF and ESD signals can be separated based on their respective frequencies as well. Frequency of RF circuits is in GHz range, while frequency of an ESD signal is tens of MHz. Therefore, a filter can be designed to short the low frequency ESD signal to ground, while passes the high-frequency RF signal. Figure 8-11 shows the input stage of a common source LNA where an inductor is added to provide ESD protection [8].

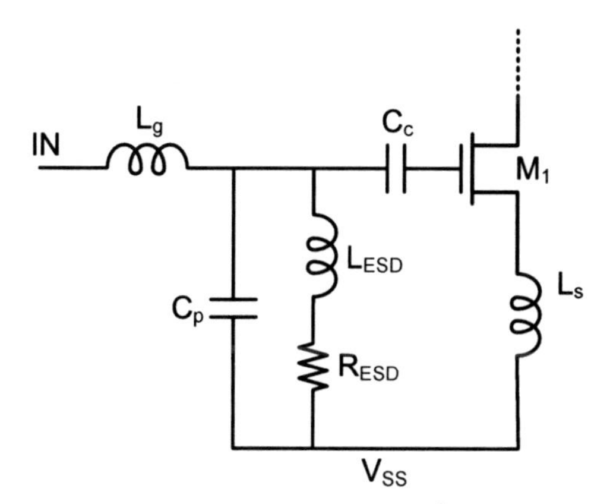

Figure 8-11. Inductive ESD protection for a common source LNA.

In this figure, C_p is the inherently present parasitic input capacitance and C_c is the coupling capacitor. ESD protection is provided with inductor L_{ESD} that has a finite series resistance R_{ESD}. RF signal sees a high pass filter, which consists of L_{ESD} and C_c and ESD signal sees a low pass filter.

First, let's consider this circuit under ESD conditions. The inductor L_{ESD}, creates a fully bidirectional path for the ESD current to ground. Hence, by adding a bidirectional clamp between V_{DD} and V_{SS}, a complete ESD protection scheme for all zapping modes can be designed. The important factor in the design of L_{ESD} is that the series resistance R_{ESD} acts as the on resistance of the ESD device. Therefore, it's crucial to minimize this resistance for high ESD protection level. In 0.18 μm CMOS technology, using an inductor with total area of 0.1 mm^2, an HBM protection level of over 3 kV was measured for the whole LNA [9].

From RF point of view, this LNA is similar to LNA with Π-type matching network where an extra parameter is available that can be used to optimize the performance of the LNA.

Using this method, a 5 GHz LNA was designed in 0.18 μm CMOS technology [9]. HBM measurement results show a protection level of over 3 kV for all four zapping modes. At the same time, the LNA shows acceptable RF performance, where –20 dB S_{11} and S_{22} was achieved with 20 dB gain and 3.5 dB noise figure.

4.4 Distributed ESD Protection

At higher frequencies, parasitics of ESD protection circuits limit bandwidth and power transfer of RF systems. For narrowband RF applications, as were the focus of this chapter so far, ESD protection can be provided using matching networks to tune out the ESD parasitic capacitance. However, in broadband applications these techniques cannot be used with success. In such applications distributed protection scheme is the preferred solution [10]. This method is based on distributed amplifier [11], where amplifiers are replaced by ESD protection devices. Consider the circuit shown in Figure 8-12(a). ESD protection circuit is modeled with a capacitor C_{ESD} which is connected to a 50 Ω source through Z_0. Z_0 provides impedance matching and is implemented using coplanar waveguide (CPW) or a simple inductor. It has been shown that dividing C_{ESD} into sections as in a transmission line allows operation at higher frequencies [10]. Figure 8-12(b) shows the *n* section transmission line structure.

Increasing the number of stages increases the cutoff frequency, which is advantageous from RF performance point of view. Now let us look at the *n*-section transmission line as an ESD protection circuit. As the number of stages is increased, size of each protection device is decreased, which limits the current flow through each device. In other words, the first devices in the current path face larger current and may cause early ESD failure. Hence, uniformity of current flow through all ESD devices becomes a concern as *n* increases. It has been shown that dividing the ESD capacitance into four stages gives an optimum performance for both RF signal and ESD stress [10]. Circuit simulations show that using a total ESD capacitance of 200 fF at 10 GHz attenuation of the RF signal is only 0.273 dB.

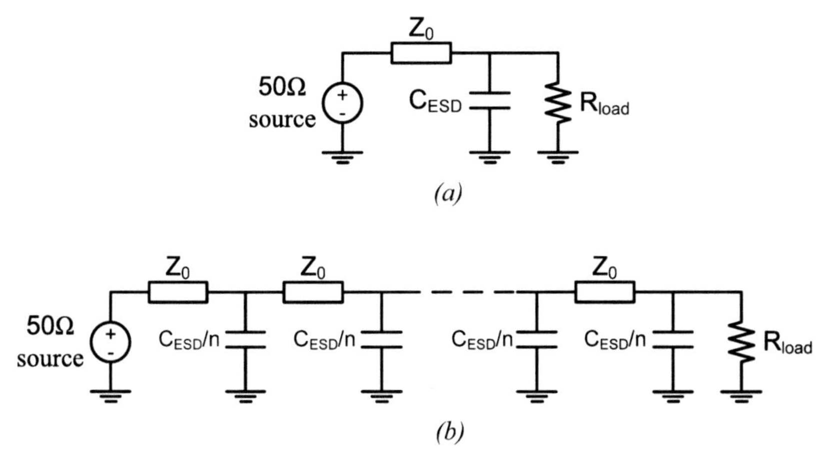

Figure 8-12. (a) Equivalent circuit with general ESD protection device, (b) n-section transmission line structure.

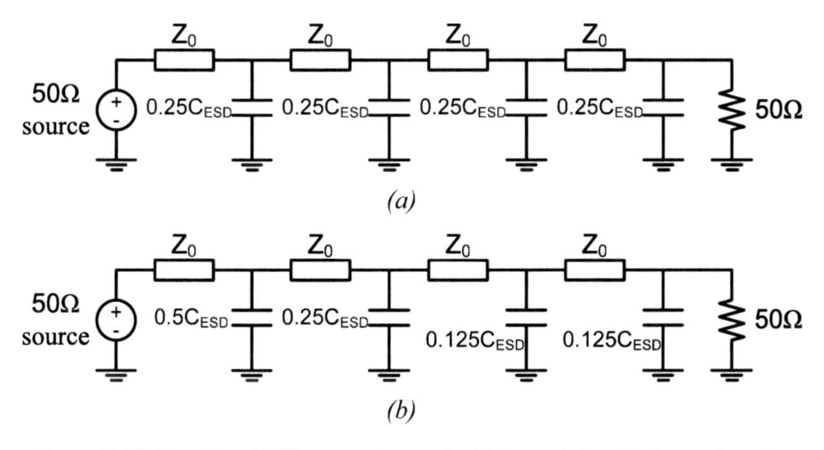

Figure 8-13. Distributed ESD protection method (a) equal-size (b) decreasing-size.

The uniform triggering of all ESD devices can be further improved by applying a non-uniform sizing method. This technique is called decreasing-size distributed ESD protection where devices that are closer to the pad are designed to be larger [12]. Figure 8-13 compares sizing of ESD devices of a four stage distributed network for equal-size (ES-DESD) and decreasing-size (DS-DESD) distributed ESD protection methods.

In order to compare these two techniques, the two distributed ESD networks were fabricated in 0.25 μm CMOS technology. Total ESD capacitance in both cases was 200 fF. Therefore, in ES-DESD all capacitances were

50 fF, while in DS-DESD capacitances were 100, 50, 25 and 25 fF. Z_0 was implemented with CPWG and was equal to 70 Ω for both cases.

RF performance of the two protection schemes was compared based on their S-parameters. Measurements were done in a wide frequency range between 1 and 15 GHz. Measured S_{11} and S_{21} are shown in Figure 8-14. Figure 8-14(a) shows S_{11} and it can be seen that S_{11} of ES-DESD and DS-DESD are almost the same in lower frequency range. As the frequency goes beyond 10 GHz, S_{11} of ES-DESD becomes a little smaller than DS-DESD. Figure 8-14(b) compares S_{22} of the two methods and it can be seen that DS-DESD has higher S_{22} compared to ES-DESD.

ESD performance is compared using HBM measurements. ES-DESD shows a HBM protection level of 5.5 kV. On the other hand, as explained earlier, DS-DESD shows much better performance where the HBM failure was over 8 kV. Overall, DS-DESD shows comparable RF performance but much better ESD performance compared to ES-DESD.

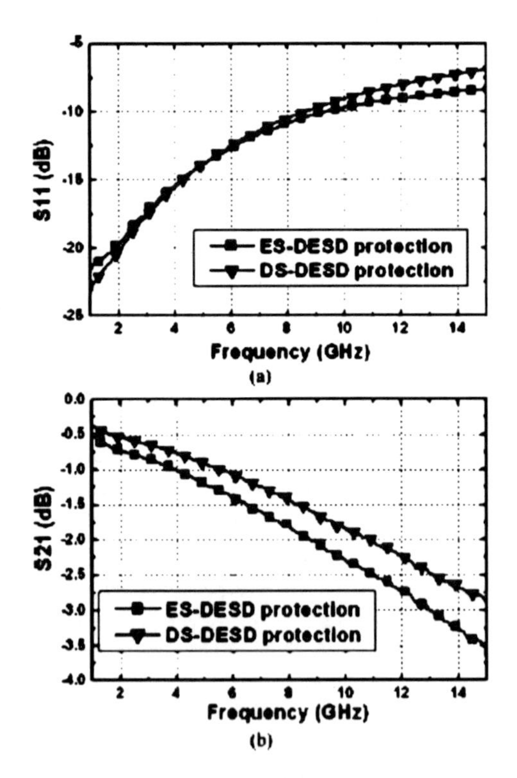

Figure 8-14. Comparing ES-DESD with DS-DESD (a) S_{11} measurement, (b) S_{22} measurement.

5. SUMMARY

In this chapter, ESD protection methods dedicated to radio frequency (RF) circuits were discussed. It was shown that ESD protection circuits for RF applications can be divided into two different categories. The first category uses regular snapback-based devices for ESD protection and the impact of the protection circuit on RF performance is suppressed using inductive networks. Examples of this category that were shown in this chapter are as follows:

1. Cancellation technique: this method uses a simple inductor to cancel capacitance of the protection circuit at the operating frequency.
2. Protection using Π-type matching network, which is an RF-ESD co-design method.
3. Distributed ESD network.

The second category differentiates RF signal and ESD stress based on their frequency. In this case, the protection circuit is a simple filter that passes RF signal but attenuates the ESD stress.

REFERENCES

[1] E. H. Armstrong, "The super-heterodyne – its origin, development, and some recent improvements," *Proc. of the IRE*, vol. 12, pp. 539–552, 1924.

[2] T. H. Lee, *The design of CMOS radio-frequency integrated circuits*, Cambridge University Press, New York, 1998.

[3] G. Gramegna, A. Magazzu, C. Sclafani, M. Paparo, and P. G. Erratico, "A 9 mW, 900 MHz CMOS LNA with 1.05 dB Noise Figure," *Proc. European Solid State Cir. Conf.*, pp. 155–158, 2000.

[4] D. J. Cassan and J. R. Long, "A 1 V 0.9 dB NF low noise amplifier for 5-6 GHz wireless LAN in 0.18 μm CMOS," *Proc. Custom Integrated Cir. Conf.*, pp. 419–422, 2002.

[5] S. Hyvonen, S. Joshi, and E. Rosenbaum, "Cancellation technique to provide ESD protection for multi-GHz RF inputs," *Electronics Letters*, vol. 39, pp. 284–286, 2003.

[6] V. Vassilev, S. Thijis, P. L. Segura, P. Wambacq, P. Leroux, G. Groeseneken, M. I. Natarajan, H. E. Maes, and M. Steyaert, "ESD-RF co-design methodology for the state of the art RF-CMOS blocks," *Microelectronics Reliability*, vol. 45, pp. 255–268, 2005.

[7] V. Vassilev, S. Jenei, G. Groeseneken, R. Venegas, S. Thijis, V. D. Heyn, M. Natarajan, M. Steyaert, and H. E. Maes, "High frequency characterization and modeling of the parasitic RC performance of two terminal ESD CMOS protection devices," *Microelectronics Reliability*, vol. 43, pp. 1011–1020, 2003.

[8] P. Leroux and M. Steyaert, "High-performance 5.2 GHz LNA with on-chip inductor to provide ESD protection," *Electronic Letters*, vol. 37, pp. 467–469, 2001.

[9] P. Leroux and M. Steyaert, "A 5 GHz CMOS low-noise amplifier with inductive ESD protection exceeding 3 kV HBM," *European Solid State Cir. Conf.*, pp. 295–298, 2004

[10] C. Ito, K. Banerjee, and R. W Dutton, "Analysis and design of distributed ESD protecttion circuits for high-speed mixed-signal and RF ICs," *IEEE Trans. on Electron Dev.*, vol. 49, No. 8, pp. 1444–1454, 2002.

[11] E. L. Ginzton, W. R. Hewlett, J. H. Jasberg, and J. D. Noe, "Distributed amplification," *Proc. of the IRE*, vol. 36, pp. 956–969, 1948.

[12] M. D. Ker and B. J. Kuo, "Decreasing-size distributed ESD protection scheme for broadband RF circuits," *IEEE Trans. on Microwave Theory and Techniques*, vol. 53, No. 2, pp. 582–589, 2005.

Chapter 9

CONCLUSION

1. INTRODUCTION

ESD has been considered as a major reliability threat in semiconductor industry for decades. As CMOS technology scales down, design of ESD protection circuits becomes more challenging. This is due to thinner gate oxides, shallower junction depths, and smaller channel lengths in advanced technologies that make them more vulnerable to ESD damages. As a result, design window for the ESD protection circuit becomes narrower. The design window is often interpreted as the voltage difference between the avalanche breakdown voltage (V_{t1}) and gate oxide breakdown voltage. For example, in 90 nm technology, the gate oxide breakdown voltage under ESD conditions is only 1 V above the drain junction breakdown voltage which leaves only 1 V design window for the ESD protection circuit. Furthermore, scaling of CMOS technology is often accompanied by increase in operating frequency, increase in total chip area, and number of packaged pins. Higher operating frequency sets a limit on the maximum parasitic capacitance of the ESD protection circuit for packaged pins. At the same time, increase in the number of pads limits the available area for ESD protection circuits. As a result, considerable effort is needed to meet the speed requirements, while satisfying the area and robustness requirements of the ESD protection circuits.

In order to design a robust ESD protection circuit, a deeper insight of the device behavior under high current and high voltage stress conditions is required. Unfortunately, traditional transistor models for circuit simulations are not well suited such an exercise. Therefore, one can investigate such a

O. Semenov et al., ESD Protection Device and Circuit Design for Advanced CMOS Technologies, 219–222.

behavior using device level simulations. The device level simulation results can be incorporated into simulations at higher levels of abstraction such as circuit and chip levels. The most important steps in the ESD design and evaluation phase are as follows:

1. Process/Device simulation and calibration
2. Mixed-Mode ESD simulation
3. Chip-level ESD simulation
4. Test chip development
5. ESD measurements.

There are two major schemes to provide ESD protection for integrated circuits: (i) snapback-based and (ii) non-snapback-based protection. Snapback-based protection method uses snapback devices to provide ESD protection between pad and V_{DD} (PD and ND modes), and pad and V_{SS} (PS and NS modes). MOSFET and SCR are the most popular devices used in this category. On the other hand, in non-snapback-based protection, each pad is connected to the supply lines with a diode and the protection for all four zapping modes is provided through these devices and a clamp between V_{DD} and V_{SS} lines.

As a snapback device, SCR has the highest protection level per unit area, which makes it the best choice in snapback-based protection scheme. The main design challenge in designing an ESD protection circuit using SCR-based devices is to reduce the first breakdown voltage and improve latch-up immunity. The most common method to reduce the first breakdown voltage is to add different triggering techniques to the main protection device. Gate-triggering and substrate triggering are two of the most common triggering techniques. In order to improve latch-up immunity, holding voltage or holding current of the SCR should be increased.

In non-snapback-based protection method the main challenge is to design a fast transient power supply clamp. This clamp should meet several requirements in an ESD event: the clamp should quickly trigger and discharge the ESD energy completely; under normal power-up conditions the clamp should remain “off”; finally, it should not oscillate under ESD stress or under power supply ramping conditions. The most common design approach divides the triggering mechanism into two sections: a rise time detector to detect the rise time of the ESD stress, and a delay element to keep the clamp “on” to discharge all the ESD energy. In addition, the clamp should be able to react quickly so that stress voltage does not go beyond acceptable voltage level for a given technology. SRAM-based and thyristor-based clamps are two examples of transient clamps.

In the design of an ESD protection circuit, in addition to ESD robustness, the interaction between the main circuit and the protection circuit should be well understood. By designing ESD protection for a pad, a non-linear parasitic capacitance is added to the pad. This extra capacitance degrades the frequency response, which is critical in high speed I/Os. Furthermore, as this capacitance is non-linear in nature, it degrades linearity of the main circuit. There are a limited number of studies in this issue in literature. The most important studies are on the impact of the protection circuit on second and third order harmonic distortions (HD2 and HD3) of analog to digital converters and on jitter of current mode logic drivers.

As the frequency of operation is increased, study of the interaction between the main circuit and the protection circuit is becoming more critical. A general method to decrease the impact of the protection circuit on the main circuit's performance is to reduce the parasitic capacitance of the protecttion circuit. However, in RF applications, a new set of ESD protection techniques are used to eliminate the impact of the parasitic capacitance of the protection circuit on the main circuit. These techniques are applicable to RF applications where the input stage is a narrowband LNA. These protection techniques can be divided into two main categories. In the first category, the impact of the parasitic capacitance of the protection circuit is suppressed at the operating frequency of the LNA using inductors. Cancellation technique and the RF-ESD co-design method are the examples of this category. In the second category, RF signal and ESD stress are separated based on their frequency difference. In other words, a proper filter is designed to create two separate paths for ESD stress and the RF input signal: RF signal is applied to the input of the LNA, while ESD stress is transferred to the ground.

2. FUTURE WORK

Although there have been significant improvements in on-chip ESD protection circuit design, there are still several unresolved issues that should be investigated. During the design phase, predicting performance of the ESD protection circuit is a major challenge. In many cases the CMOS process needs to be characterized to be able to predict the performance of the protection circuit. A hit and miss, iterative approach may require too much time and resources. It can be seen that there is a need to develop accurate simulation tools to predict ESD performance with a great degree of confidence at all levels of abstraction including the chip level. Moreover, the accuracy of ESD models for simulations is extremely important in order to predict the ESD circuit performance with confidence. These models can be used by circuit and system designers to predict the impact of ESD protection on the main circuit's performance.

As CMOS technology is scaled, the operating frequency of circuits is increasing. Hence, the impact of ESD protection circuit on very high speed circuits with data rates exceeding 10 Gbps should be investigated. As a result, it's necessary to focus on reducing parasitic capacitance of ESD protection circuits. FOM1' is the figure of merit that should be maximized and is defined as follows:

$$FOM1' = \frac{V_{HBM}}{C_{ESD}}$$

Furthermore, as modern chips consist of analog and digital blocks with multiple supplies and ground pins, a proper ESD protection scheme between these pins is becoming very challenging. These ESD protection circuits should be able to provide the required protection in addition to proper isolation between analog and digital blocks. This issue is more critical in ground pins where the noise coming from the digital block should be completely isolated from the analog block.

Finally, providing protection against CDM stress is becoming more challenging in today's semiconductor industry. There are a number of factors that lead to these challenges: First of all, occurrence of CDM stresses are increasing as degree of automation is increased in handling of chips. Secondly, in advanced technologies the damages (gate oxide, junction breakdown) will occur at lower voltages. Thirdly, relatively slow ESD protection circuits primarily designed for HBM are not able to trigger quickly enough to dissipate the ESD energy associated with the CDM stress. Finally, larger, faster chips require complex packages that have higher capacitances for decoupling purposes, which results in higher CDM discharge currents through the device. Therefore, it is necessary to improve the turn-on speed of ESD protection devices and provide better immunity to CDM stress.

INDEX

A

B

C

D

E

F

G

H

I

J

K

L

M

T

U

V

Z

Printed in the United Kingdom
by Lightning Source UK Ltd.
135940UK00007B/321/P